# In Quest of
# Tomorrow's Medicines

**Springer**
*New York*
*Berlin*
*Heidelberg*
*Hong Kong*
*London*
*Milan*
*Paris*
*Tokyo*

Jürgen Drews

# In Quest of Tomorrow's Medicines

Translated from the German by David Kramer

With 38 Figures

Springer

Jürgen Drews
Orbimed Advisors
767 Third Avenue
New York, NY 10017-2023
USA
*and*
International Biomedicine
  Management Partners
CH-4010 Basel
Switzerland

*Cover illustration:* Tom Tracy/Stock Market

Library of Congress Cataloging-in-Publication Data
Drews, Jürgen, 1933–
    In quest of tomorrow's medicines/Jürgen Drews.
      p.  cm.
    Includes bibliographical references.
    ISBN 0-387-95542-9 (pbk.: alk. paper)
    1. Drugs—Research.   2. Pharmaceutical industry.
 3. Pharmaceutical technology.   I. Title.
    [DNLM: 1. Drug Industry.   2. Drug Approval.   3. Technology
 Pharmaceutical.   4. Research.   QV 736D776i   1998]
    RS122.D74   1998
    615'.19—dc21                                              98-27801

ISBN 0-387-95542-9   (softcover)              Printed on acid-free paper.

Printed in the United States of America.

9 8 7 6 5 4 3 2 1          SPIN 10885703

www.springer-ny.com

Springer-Verlag   New York Berlin Heidelberg
A member of BertelsmannSpringer Science+Business Media GmbH

# Contents

Figure 1.1.  Dispensary, nineteenth century. (Source: *Die Apotheke. Historische Streiflichter*. 1996, Editiones Roche. F. Hoffmann–La Roche AG, Basel, p. 123. Universitätsbibliothek Basel, Portraitsammlung.)

# 1
〜〜〜
# Drugs, Patients, and the Pharmaceutical Industry

## The Role of Drugs in the Practice of Medicine

In thinking about the pervasiveness of technology in our lives as the twentieth century draws to a close, it perhaps does not immediately occur to us that pharmaceutical products play as integral and important a role in our civilization as other achievements of ours: airplanes, automobiles, and other means of transportation, for example; or radio, television, newspapers—in short, all the means of communication that accompany us throughout our days.

It is already a commonplace that people respond to minor fluctuations in their health by taking pills—pills for headache and for toothache; sleeping pills and tranquilizers; pills to lower the temperature, quiet the cough, and clear the sinuses when one has a cold; medicines to reduce the appetite; preparations against diarrhea and constipation, against heartburn, against nausea. We could continue for many pages with a list of medicines taken for such everyday complaints. In this we would be describing only one aspect of the use of pharmaceuticals, and a rather trivial one at that. In the war against serious disease, medicines are an indispensable weapon in the physician's arsenal. In this realm medicines save lives, or at least prolong them and make them more bearable. We shall have much more to say about the effectiveness of medicines in the course of the book.

Despite the central role that pharmaceuticals play in the current practice of medicine and, more generally, in our civilization, few people know where medicines come from or how the pharmaceutical industry discovers them and develops new products. Furthermore, most people have only the vaguest notion of the criteria by which the effectiveness and safety of modern medicines are measured, and they know almost nothing about the research that has produced our modern pharmacological treasure-trove. A lack of knowledge about drugs almost inevitably has led to a lack of understanding of the societal roles of those institutions that have been the main providers of medicines during the last century. This book represents an attempt to redress this deficiency.

## Pharmaceuticals as "Innovation"

Pharmaceutical research as we know it today is still a young field. It developed somewhat over a century ago when several scientific disciplines arrived almost simultaneously at a stage of development that made possible the practical application of their various achievements. These disciplines included analytical chemistry, synthetic chemistry, and experimental pharmacology. Whenever a sudden leap occurs in the development of a scientific discipline, there arises the opportunity for innovation. This is the case particularly when the respective disciplines complement one another.

Today, the term "innovation" is applied in industry and politics to any technical or scientific development. The term is used so broadly that it has practically lost all meaning. In the classic definition of the Austrian-born American economist Joseph Schumpeter, however, this word takes on a concrete meaning, namely, a new product or process at the moment it is introduced to the market.[1] New drugs or therapeutic procedures would be considered innovations according to this definition. That such innovations suddenly became possible at the end of the nineteenth century in considerable numbers was due to the fact that at this time a modern pharmaceutical research program had developed out of chemistry and experimental pharmacology.

And furthermore, this new scientific discipline, whose raison d'être was the development and classification of new active agents, established an institutional basis for itself in the newly developing pharmaceutical industry. The turn of the century was a time when the burgeoning sciences and their associated industrial applications were presented with a favorable climate—an *innovative* climate, as many of today's managers would put it.

In this book we shall trace the developments out of which modern pharmacological research arose. We shall also describe the historical, economic, and institutional bases of pharmaceutical innovation. Again today, a hundred years later, we live in a time of radical scientific, institutional, and economic change. In relation to this we must ask about the future of pharmacological research and associated institutions, and we shall see in the course of our investigation that drug research has been nourished and enriched by enormous technical advances. The newest of these technical revolutions, molecular biology, is about to provide a new conceptual and technical framework for biology and medicine.

Pharmaceutical research will not be exempt from this revolution. For a hundred years the search for new medicines and their development has been the concern of the great pharmaceutical firms, and indeed, the pharmaceutical industry has possessed a virtual monopoly on the discovery, development, production, and distribution of medicines. Now other players have appeared upon the scene: small, highly technical firms that arose out of molecular biology, but that came also from many fields of chemistry and pharmacology, as well as contract research organizations, firms, that is, that are concerned with particular segments of the processes of medical research and development. Thus we see the situation that many universities, at least in the United States, are ready to involve themselves in the fundamental aspects of pharmaceutical research. The deck is clearly being reshuffled. Just as it was a hundred years ago, the recipe for success will come not from doubt and hesitation, but from posing and solving the great problems of biomedical research, from the intelligent transfer of scientific discoveries into new products and processes, that is, into "innovations," and in the sensitive reaction to social needs throughout the world.

For like other human undertakings, the development of medicines and how they are used are matters not of national, but of global, concern. The contribution of an individual, of individual institutions, of individual firms or nations, will be judged according to its usefulness for a large portion of humanity, if not for all of us.

*Note:* Medical terms employed in the following sections will be explained either directly in the text or in the Glossary.

## Medicines Achieve Cures . . .

Modern medicine has been greatly influenced by modern science, with the result that today drugs play a central role in medical practice, which was not the case in earlier epochs. In earlier medical cultures, drugs or druglike preparations often played only a marginal role in a more comprehensive program of treatment, one that aimed at restoring the patient to a harmonious state—that is to say, a complete psychosocial restoration.

Here drugs played a supporting role, never a leading one. This situation has changed in many areas of modern medicine. The treatment of infections, for example, is unthinkable without chemical agents and antibiotics. To treat a lung inflammation caused by a pneumococcus, an erysipelas, a bacteriological inflammation of the middle ear, or a case of peritonitis *without* antibiotics would today be considered medical malpractice. The same can be said for the treatment of many fungal infections with antifungal agents and to an increasing degree for the treatment of viral infections with antiviral agents. To withhold treatment in the form of a combination of reverse-transcriptase inhibitors and a protease blocker from a patient with HIV (human immunodeficiency virus) would likewise be considered malpractice.

And however much physicians, especially the older ones among us, are convinced of the synergy of all medical measures, at whose center the patient stands as unique and distinctive personality, it has nonetheless become clear how often medicines play a decisive role, as some examples here presented demonstrate. This refers as much to the selection of the correct drug or combination of drugs as to their correct use. In the treatment of many infections and parasitic infestations, antibiotics and chemical agents are absolute measures, "absolute" in the sense that their use may be accompanied by other measures but can never be replaced by them (Figure 1.2).

A glance at other types of therapies makes it clear that drugs play an almost universal role in modern medicine. To be sure, the effective treatment of hypertension and its arteriosclerotic complications should be treatable through weight control, dietary measures such as reduction in salt intake, sufficient sleep, avoidance of nicotine and excessive consumption of alcohol, and exercise. Nonetheless, modern drugs such as beta-blockers, calcium antagonists, centrally acting antihypertensive drugs, angiotensin converting enzyme (ACE) inhibitors—to name just the most important—are necessary components, perhaps even the fundamental component, of the treatment (see Table 1.1). Without these substances the overall mortality from circulatory disease, which has been declining in the industrialized world, would again be on the rise.

Thirty years ago, gastric and duodenal ulcers were the occasion for extensive surgical intervention. Today they are largely medicinally controlled. H2 histamine blockers such as cimetidine and ranitidine have brought about a significant reduction in the costs associated with these diseases. A 1980 report of the German Association of Pharmaceutical In-

**Reduction in Mortality for Several Bacterial Infections Through Antibiotics**

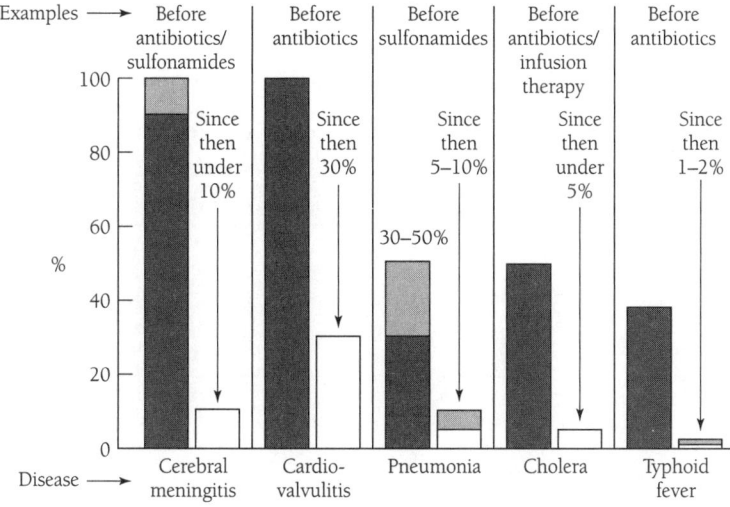

Figure 1.2.   Mortality as a percentage of patients: Influence of antibacterial therapies.

dustries contains the information that in that year, in Germany, 400,000 patients were treated for duodenal ulcer. Of these, 59,000 had to be hospitalized. The cost to the federal economy for cases of duodenal ulcer was about 943.5 million DM (deutsche marks). Of this, 514 million DM was attributed to lost productivity. The cost of treatment, 430 million DM, was about equally divided between those who were treated as outpatients and those who were hospitalized. Comparing this with 1977, the year in which cimetidine became available in Germany, by 1980 a savings of 200 million DM in treatment costs had been achieved. If we subtract the cost of the medication, 43.1 DM, then there remains a "net profit" to the national economy of 150 million DM.[2] The situation in 1984 was even better (see Figure 1.3).

Today, after the discovery of the bacterium *Heliobacter pylori* as a significant factor associated with ulcers, duodenal ulcers can be "cured" through medicinal treatment. And it is not only acute ulcers that can be healed; by eliminating *Heliobacter pylori*, the incidence of recurrence has been greatly reduced. Many acute illnesses such as coronary thrombosis and strokes caused by blood clots are, when detected early, so treatable with thrombolysis that long-term damage can be avoided or at least greatly limited. Perhaps we should say a word here about the role of painkillers, which, even if

Table 1.1.    Important new drugs: The first letter of the name of a drug is located on the time scale corresponding to its first introduction.

**Improved Treatment Due to Innovative Drugs**
*Selected Examples of Treatments*

| | <1960 | 1960 | 1970 | 1980 | 1990 | 2000 |
|---|---|---|---|---|---|---|
| Trans-plantation | | | | Aza-thioprine | Cyclosporin A | Mycophenolate mofetil, Tacrolimus, mAB monoclonal antibodies |
| Heart attack/ coronary heart disease | Heparin | | Streptokinase, beta-blockers | | t-PA, ASS, ACE-inhibitors | |
| Hyper-tension | | | Beta-blockers    Calcium antagonists | | ACE-inhibitors | Angiotensin II receptor antagonists |
| Gastro-intestinal ulcer | | | | H$_2$-blockers | | Proton pump inhibitors, antibiotics |
| Cancer | | | Vincristine    Mitomycin    Chlorambucil    Methotrexat | Cisplatin | Alpha interferon, Interleukin-2 | Paclitaxel, mAK    Retinoids    G-CSF (therapy supporting) |
| AIDS | | | | | HIV reverse transcriptase inhibitors | HIV protease inhibitors |
| Bacterial infections | | Semi-synthetic penicillins    Cotrim-oxazol | Cephalosporine, Rifampicin, Vancomycin | | Fluoroquinolones    Carbapenemes | |
| Year ⟶ | | 1960 | 1970 | 1980 | 1990 | 2000 |

they are not necessary to preserve life, are nonetheless of great importance in improving the quality of life. There is hardly a person in the industrialized world who has not at one time or another found such drugs an enormous blessing.

## . . . and Facilitate Other Forms of Treatment

Medicines are important not only in their direct effects against illnesses, but in their secondary role of supporting other therapeutic techniques or even making them possible in the first place. All areas of surgery rely on anesthesia and, for the most part, also on muscle relaxants. Without a large pool of medications on which to draw that make possible controlled

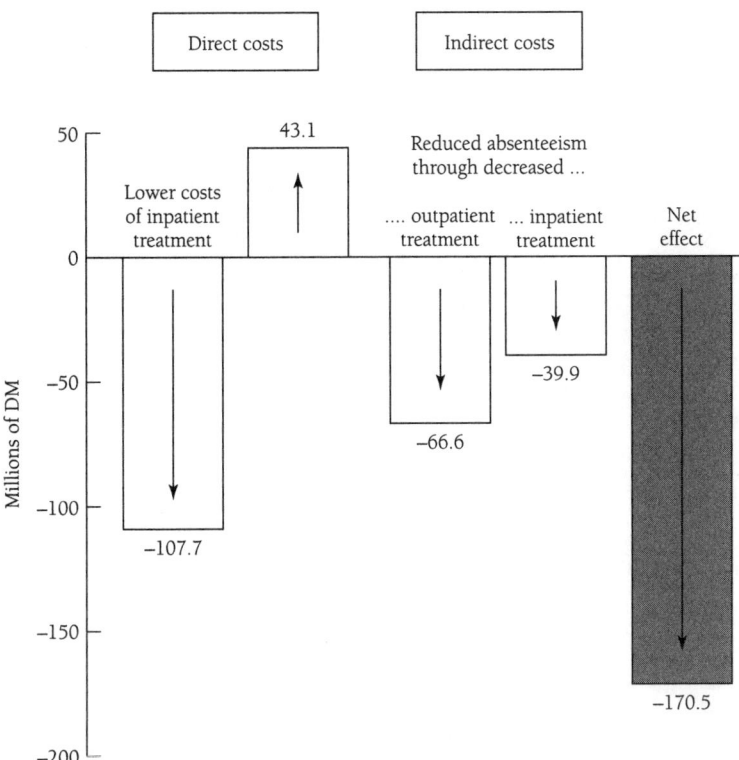

Savings Through $H_2$ Blockers in Germany

Figure 1.3.    The arrows pointing downward indicate cost savings; those pointing upward indicate additional expenses. Total savings in 1984 amounted to 170 million German marks.

anesthesia and muscle relaxation, there would be no modern surgery. Without immunosuppressants there would be no transplant surgery, to give one more example from the surgical domain. If steroids, cyclosporine, azathioprin, mycophenolate, and antisera against T-cells were suddenly no longer available due to some catastrophe, then there could be no more organ transplantations. The failure of an organ, the liver, say, would then be fatal. Even chronic kidney failure would be incompatible with a "normal" life, as is evident from the example of those who must undergo continual dialysis.

Psychopharmacology has created a revolution in psychiatry, a fact that at least younger doctors are no longer aware of. If psychiatric institutions before the discovery and development of antipsychotics, antidepressants, and tranquilizers were often in reality macabre places of madness, terror, and the use of force, they have come more and more to resemble, since the introduction of phenothiazines and other neuroleptics, clinics for internal medicine. Today there is no longer much use for such blunt and restrictive measures as straitjackets and isolation cells, and the more drastic forms of therapy like electroshock play only a marginal role. These psychotropic medicines are important not only for their management of the symptoms of illness. They also facilitate the interaction between doctor and patient, thus creating an environment in which psychotherapy can effectively take place (see Table 1.2).

## Drugs as Diagnostic Tools

Furthermore, drugs and other substances that satisfy the criteria of drugs with respect to effectiveness and, above all, safety are important components of a diagnostic regime. We may give as examples x-ray contrast

**Table 1.2.**   This table lists only the most important new medicines.

### Important Substances Since the Mid-1980s

| | |
|---|---|
| Substances against cardiovascular diseases | • HMG-CoA-reductase inhibitors (lipid reducers)<br>• ACE-inhibitors (extension of indications to cardiac insufficiency)<br>• Angiotensin-II receptor antagonists<br>• t-PA (tissue plasminogen activator)<br>• T-calcium channel blockers<br>• Endothelin antagonists |
| Substances against cancer | • Cytokines (IFN-alpha, IL-2)<br>• Paclitaxel<br>• Retinoids<br>• Humanized monoclonal antibodies |
| Substances against AIDS and consequent diseases | • HIV reverse transcriptase inhibitors<br>• HIV proteinase inhibitors<br>• Fluconazole<br>• Itraconazole (systemic fungal infections)<br>• Ganciclovir (CMV infections) |
| Vaccines | • Hepatitis A and Hepatitis B<br>• Meningitis through Hemophilus influenzae b |
| Other significant new introductions | • Erythropoietin (Anemia in dialysis patients)<br>• G-CSF, GM-CSF (neutropenias, etc.)<br>• Ondansetron (antiemetic)<br>• Interferon v-beta (multiple sclerosis)<br>• Virally inactivated blood products<br>• Recombinant blood products (factor VII/IX)<br>• Acarbose (diabetes mellitus)<br>• Lipstatin (fat resorption)<br>• Mycophenolate, tacrolimus (transplantations)<br>• Selective serotonin uptake inhibitors<br>• Selective monoaminooxydase inhibitors (depression) |

media for imaging blood vessels and internal organs (kidney, pelvis of the kidney, gall bladder, etc.), new organometallic compounds in the operation of modern imaging techniques such as positron emission tomography (PET), and organic iodine compounds in CAT (computer-aided tomography) scans. These methods make possible the depiction of organs and tissues with images whose resolution and clarity far exceed what is attainable with conventional x-rays.

Beyond their role in medicine—that is, diagnosis and treatment—drugs play an important role in society at large. Drugs that have been proven effective and essentially harmless are sold without prescription or a physician's oversight and are taken without much thought by large numbers of people for such everyday complaints as headaches, colds, menstrual discomfort, and muscular pain.

The list could be extended effortlessly, but we would not thereby be saying much that was new. It was not our intention to give the impression here that medicines are the solution for all medical problems or indeed for all the problems of life. We wished simply to sketch briefly that modern medicine in its diagnostic and especially its therapeutic dimensions is unimaginable without drugs. Drugs are not everything in medicine, but there would be no modern medicine without them (see Table 1.3).

## Illness as Malfunction: Treatment as Repair

The liberal use of drugs—in large part still physician-controlled but to a growing extent neither medically supervised nor approved—reflects a particular attitude toward health and the body. Sickness is classified as a rectifiable and reparable malfunction and not as an integral part of one's personal fate, a fate that in its various manifestations may constitute a life-long struggle. Here one cannot help but think of the analogy between medical practice or the management of a hospital and an auto repair shop, in which parts are replaced or repaired, and the car's proper functioning is restored to the owner's satisfaction. The customer pays, and for his money he demands definite results of a guaranteed quality.

Should these results fail to materialize or should the quality not be up to the customer's expectations, there is a process for settling complaints that extends all the way to legal proceedings. Especially in the United States one sees a willingness to blame the treatment process for negative outcomes of treatment and for actual, or even only imagined, injury.[3] And of course, large practices and hospitals are favored targets for accusations and lawsuits—because they are emotionally and politically easy—even

Table 1.3.  Diagnostics serve in recognizing diseases and in monitoring the course of treatment. The simpler tests can be administered by the patient.

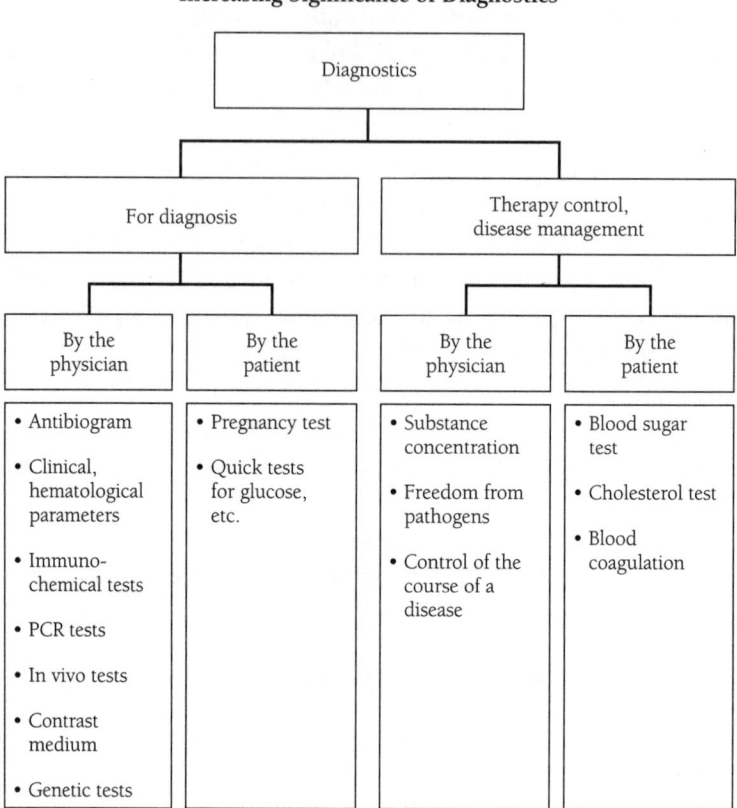

**Increasing Significance of Diagnostics**

| Diagnostics | | | |
|---|---|---|---|
| **For diagnosis** | | **Therapy control, disease management** | |
| By the physician | By the patient | By the physician | By the patient |
| • Antibiogram<br><br>• Clinical, hematological parameters<br><br>• Immuno-chemical tests<br><br>• PCR tests<br><br>• In vivo tests<br><br>• Contrast medium<br><br>• Genetic tests | • Pregnancy test<br><br>• Quick tests for glucose, etc. | • Substance concentration<br><br>• Freedom from pathogens<br><br>• Control of the course of a disease | • Blood sugar test<br><br>• Cholesterol test<br><br>• Blood coagulation |

when a failure to maintain accepted standards of treatment cannot be proved.

Medicines are viewed by a consumer-oriented society as products that one takes with certain expectations as to their efficacy. That this attitude is so much more widespread in the United States than in Europe may have something to do with the fact that Americans, to a much greater degree than most Europeans, pay for their own drugs.

Thus medicines have played a very important role in the secularization of medicine, in uncoupling the medical content from a metaphysical and religious understanding of human existence. How greatly the belief in the manipulability of the human body—the utilitarian view that its functioning can be preserved by medicines—has taken hold in our society can be seen clearly in the case of vitamins. The important role of these substances as "biocatalysts" was described in great detail in the first half of the twentieth century, and some of these substances, for example the D vitamins and the retinoids (derivatives of vitamin A), are still of interest to researchers on account of their biological and pharmacological functions in the differentiation of cells and tissues.

The nutritional–physiological role of these substances has likewise been adequately described, and for almost all vitamins there exist "recommended daily allowances." Today, the intake in a balanced diet of necessary amounts of vitamins is no longer a problem—not, at least, in the industrial nations, in which food manufacturers in many cases enrich their products (milk, flour products, fruit juices, and others) with vitamins. Nonetheless, the regular ingesting of additional vitamins has become for many a cherished custom, even when the medical utility of such "overdosing" has not been established and at best can be derived from theoretical considerations.

Of course, there are counterexamples. Aspirin, for instance, prevents heart attacks. But this is a measure of proven effectiveness. Many allegedly "bioactive" substances, most of them essentially placebos, almost all without adequate proof of safety or efficacy, are taken for purposes of self-treatment or prophylaxis. Such purposes are reminiscent of the care that many drivers devote to their automobiles. One has a lot of money invested in the machine that is one's own body. It is irreplaceable, and so even if the vitamins do no good, at least they can't do much harm. Thus one supplies one's machine with anything that might be of use or that might be effective against self-inflicted injury (alcohol, nicotine, lack of exercise).

Even more questionable than such "internal body care" with drugs and druglike substances is the use of medicaments for purposes that have nothing to do with the cure or amelioration of disease. Hormonal contraceptives belong to this category, though at least they serve a purpose in family planning and in giving women control over their bodies. Moreover, the level of risk can be measured quite well, since such large numbers of consumers of these products are involved. All the data point to a somewhat elevated risk of thrombosis among women who take hormonal contraceptives. However, this risk is apparently offset by a decreased risk—depending on the particular contraceptive regime—of other illnesses (breast cancer, osteoporosis).[4] Furthermore, one must set the tiny risk, in absolute numbers, of thrombosis against the heavy personal, economic, and societal burdens that can result from an unwanted pregnancy.

Such qualifications are not appropriate in the case of hormones taken by athletes to increase their muscular strength. The use of anabolic steroids to add muscle mass, the taking of amphetamines to improve reaction time or to induce wakefulness, these are unambiguous examples of the abuse of medicines, even though the dangers are played down by athletic associations and occasionally even by sports doctors.[5]

We could go on with even more flagrant examples of the misuse of drugs. Those who take certain benzodiazepines, the active ingredients of many sedatives, either alone or in combination with alcohol, in order to lower inhibitions, cause altered mental states, or induce in a partner the loss of self-control with partial loss of memory, such people are engaged in criminal activity, whether with respect to themselves or to another. The fact that medicines can be abused speaks not against the medicine itself, but against the individual drug abuser and perhaps also against the society that allows such behavior and is incapable of effectively preventing it.[6]

Indeed, there is no progress in civilization that cannot somehow be abused. It would hardly make sense to do without some valuable drug just because a few individuals find dubious, sometimes even criminal, uses for it. Nonetheless, the manufacturers of drugs and the doctors who prescribe them are obligated to do whatever is in their power to prevent the inappropriate or illegal use of the products they sell or prescribe.

## The Role of the Pharmaceutical Industry in the Medical Use of Drugs

Drugs are used to cure or ameliorate disease. Throughout human history people have sought effective means of combating their illnesses and afflictions. As societies organized themselves into units based on the division of labor, this function fell more and more to physicians, and later to apothecaries. The physician wrote out a prescription, and the apothecary then prepared the medicines according to appropriate recipes. That both of these, the physician and the apothecary, billed their customers for services rendered was as a matter of principle never the subject of controversy. They offered certain services that they were qualified to perform as a result of years of study and training, and for this they were compensated. The size of the bill was a function of the scope and complexity of the services performed. Not surprisingly, payments to physicians have often been viewed as excessive. But if a society ensures that no one has to do without medically necessary treatment such as an expensive drug or surgical intervention simply because the patient can't afford it, then the conflict over fees is blunted. It becomes a technical question. In most of the industrial countries there appears to be agreement on two things: first, that a physician must (should) be paid for his services, and second, that no one should be denied critical treatment on economic grounds.

Medical luxuries, however one wishes to define them, can be reserved for those able and willing to pay for them. Essential medical services, that is, all that averts death, disability, or serious pain, must absolutely be made available to all. Something like this, fashioned according to some least common denominator, could be considered the medical credo of modern industrial societies, characterized by a religious or social ethos, or a combination of both.

At this point it is necessary to speak about the vastly different ways that this basic set of convictions has been interpreted in the various systems of health care. We should mention here that the United States has experienced, and continues to experience, difficulties in bringing a substantial portion of its less-well-off citizenry (about 35 million out of a population of 280 million) into the system of organized health care. The Medicaid program functions reasonably well in its treatment of emergencies, but it has many shortcomings in its handling of basic care among the poorer segments of society. Conversely, many European countries have taken precautions to avoid such inequalities.

What accounts for this difference? Is it a difference in value systems, or is it a problem of management? Probably both of these factors play a role. The majority of Americans would agree on purely religious and ethical grounds that poor people, those in poverty through no fault of their own, and those in need of assistance should be treated either at no cost or at a cost commensurate with their economic capabilities. The entrepreneurial–capitalistic orientation of American culture ("the pursuit of happiness" instead of the European *fraternité*) has prevented the formation of appropriate instruments for translating the existing ethos into a corresponding system of health care

## Medical Needs and Economic Forces

How do today's drugs fit into these portraits of society? Or more importantly: To what extent do the goals of modern pharmaceutical research still correspond to society's therapeutic needs? Modern drugs can be manufactured only by means of complex, multistage research and development processes. Scientists and physicians are involved in both research and development, and they represent quite different disciplines. This interdisciplinarity is at present achievable within the pharmaceutical industry. There is no other institution that unites under one roof all the functions necessary for discovery, development, and manufacture of a drug or medicine.

These pharmaceutical concerns, however, are for-profit enterprises, publicly traded companies for the most part, obligated to their shareholders. That is, they must make a profit in order to compensate those with capital invested in them. Thus, the discovery and development of new drugs, through their being embedded in industry, are thus obligated to serve not exclusively medical and scientific criteria, but economic efficiency as well. To put it more simply: In such an environment one must search out and develop only those drugs that "pay off."

We are thus dealing with two groups of criteria that today's drugs have to satisfy: scientific and economic. In the first category belong the criteria of effectiveness and safety. Ineffective medicines whose degree of toxicity stands in poor relationship to their therapeutic value have scarcely a chance of being approved in the industrialized world today. Secondly, and now we enter into the realm of economic criteria, the medicine in question must be effective against a disease with a high degree of incidence in the population: In this category belong cardiac and circulatory diseases, bronchial asthma, osteoporosis, various cancers, primary rheumatoid arthritis, and other auto-immune disorders. The size of the target population plays a decisive role in the decision for or against the development of a drug.

## Medical Orphans . . .

Rare diseases, the so-called orphan diseases, are rarely mentioned any more in the large pharmaceutical firms. Orphans are those diseases whose prevalence in the population of the United States is no more than 200,000 cases. This corresponds to a frequency of 0.91 cases per thousand of population.[7] Occasionally, a biotechnology firm develops a remedy for such a disease. Pulmozyme®, a preparation of human DNase obtained through genetic engineering, developed by the firm Genentech against cystic fibrosis, a disease of the respiratory tracts and pancreas that left untreated leads to early death, is an important and positive example. DNase is an enzyme that breaks down deoxyribonucleic acid (DNA). Aqueous solutions of DNA are extremely viscous, and victims of cystic fibrosis have high concentrations of DNA from dead leukocytes in their bronchial mucus, which consequently becomes viscous and difficult to expectorate. DNase can thin the mucus and thereby free up the respiratory tracts.

If such drugs are, for the most part, not actively searched for (as was Pulmozyme®), then intelligent legislation can nevertheless encourage the development of substances discovered in other research programs for use against rare diseases. The American "orphan drug" law allows firms to take a tax deduction for the development costs of orphan drugs and gives the firm that first introduces such a medicine exclusive rights in the American market for seven years, independent of the patent process. This law, which unfortunately has no counterpart in Europe, at any rate represents for the American market an encouragement that somewhat lessens the problem we have been describing.[8]

## . . . Orphaned Medicine

Tropical diseases like malaria, schistosomiasis, filariasis, and leishmaniasis kill millions of people every year. The medical need, about which the industry is always talking, is thus enormous. Yet every one of the large pharmaceutical companies has either abandoned its research program for tropical diseases or greatly reduced it. These diseases occur primarily or even exclusively in the Third and Fourth Worlds, which are either too poor to purchase expensive medicines or else have set economic priorities that provide no money for medicaments, especially for expensive drugs. The result? High therapeutic needs that cannot be transformed into functioning markets and that therefore are (must therefore be?) largely ignored by the pharmaceutical industry, even though the scientific conditions for discovering new substances to fight disease were never better than they are today.[9]

Sometimes it is economic or technical difficulties that stand in the way of setting certain goals. As it became clear that chemotherapies for most forms of cancer would develop slowly and in small increments, to just such an extent the pharmaceutical firms withdrew from this arena, leaving the "specialists" who were already involved in this area to develop it further.[10]

The manufacturing costs of a new substance can also pose an obstacle to its development, even if the substance promises unquestioned therapeutic advantages. Often it is the complexity of a synthesis or of a biotechnological process that is the cause of high manufacturing costs. But expensive source materials can also contribute to the problem. In this connection the manufacture of the anticancer drug Taxol® comes to mind. To obtain the drug, large quantities of the bark of a certain type of yew tree must be obtained.

When a large pharmaceutical company was developing the first protease inhibitor against AIDS, which had been discovered at a research center in England, there was a long period in which there was great doubt as to whether an economically feasible synthesis would ever be developed. A very expensive synthesis can be cost-effective only if the substance is administered in small dosages, so that any one individual needs only a small quantity. But in this case the anticipated dosage wasn't in the range of a few milligrams per day, but in the neighborhood of one to two grams! At first it seemed as though the substance would be brought to market, if at all, at a dosage of 1.2 grams per day. Concern about the economic viability of this medicine reached the point where a highly placed employee of the firm attempted to forbid the use of high dosages in the clinical trials, though from a scientific standpoint these were deemed absolutely essential. He was convinced that this medicine would forever remain unprofitable in dosages as high as 1.2 grams, later 1.8 grams. "How effective it is at higher dosages I have absolutely no desire to know," he declared.

Fortunately, the story doesn't end here. The synthesis was improved and the cost appreciably lowered; moreover, the drug was found to be effective and practically free of side effects. To the extent that the pharmaceutical industry finds itself under great cost pressures, there are more and more cases where valuable research directions are not followed up because of doubt about the profitability of the resulting medicines. A new type of substance to enlarge the bronchial tubes is removed from the development portfolio because its manufacture is too costly, thereby casting doubt on the economic viability of the medicine. A new immunosuppressant antibody comes under fire because its use could endanger the sales of another immunosuppressant that is being marketed by the same company. Cyclosporine, a standard medicine in immunosuppressive therapy today, was in danger of being removed from the research portfolio in the early stages of its development. Why? Development was expensive, more expensive than the development of drugs for migraine and venous stasis, and the marketing department, with its notorious blindness to future developments, had estimated sales for this substance that amounted to only about ten percent of what was later actually sold.[11]

The list of similar examples could be extended indefinitely. It is not a matter of finding fault with particular firms or individuals. The fundamental question to be answered is this: To what extent and at what stage of development should economic considerations be allowed to influence the research and development process? As long as the search for new drugs, and above all, their development, is almost exclusively the province of profit-oriented enterprises, it will be impossible to untangle the relationship between economic calculation and the needs of medicine.

## The Future Is "Open"

One can only do a better or worse job of dealing with the situation. An intelligent strategy must always take into account that the future is new and open and never represents a mere extrapolation of the past. A very new type of medicament like the above-mentioned cyclosporine had no market while it was under development. It created a market for itself by opening up new therapeutic possibilities. To judge the future prospects of such a drug by the sales of its rudimentary predecessors is simply wrong. Medicaments can change the way medicine is practiced, and the more a drug represents a departure from what has gone before, the greater is its chance to effect such a change.

To anticipate such change and to interpret the new drug's chances in this—still hypothetical—new framework would be the proper intellectual basis for judging a new substance and its chances of success. But there are certain prerequisites for making such extrapolations, namely, medical or scientific knowledge, a feeling for social and political developments that might affect health care, imagination, and the courage to take risks. These characteristics are often in short supply in the leading management groups of pharmaceutical firms. They are much more likely to be found in the small or medium-sized genetic engineering firms. The prospects of a single pharmaceutical firm, and, of course, those for the industry as a whole, depend on correct decisions being made at the top, and here "correct" means, "Don't forget about the future!"

## The Voice of the Patients: The AIDS Lobby

But even if this were the case, not all problems would be solved. The competitive mechanisms within the pharmaceutical industry and the cost-cutting measures taken by governments in the industrialized world favor above all the development of medicines against very common diseases. The industry requires large sales and profits in order to justify high research and development costs. Governments are concerned that medicines, including the newest drugs, not be too expensive. Price restrictions, on the one hand, and the requirements of capital, on the other, can be reconciled only when large quantities of medicines can be sold. This is possible only in the case of common diseases and assumes, moreover, that the manufacturing cost of new medications is low. In the political and economic climate currently in effect, industry will attempt to find new types of medications against common diseases that are inexpensive to manufacture.

In the long term, many diseases do not get sufficient attention. The result is that certain groups of patients feel badly treated by the pharmaceutical industry and by the health-care system in general. When such feelings of inadequate treatment attain a certain intensity, they can lead to critical reactions. The aggressive lobbying by AIDS patients—who in many countries have aligned themselves with other interest groups, some of them international, and have brought pressure on pharmaceutical firms and national governments—is a good example of such developments. In the first phase of their existence the groups Act Up, the Gay Men's Initiative, and many others relied on irrational arguments and demonstrations, leading sometimes to the use of force.

They achieved a certain success with public opinion, governments, and especially regulatory agencies like the American Food and Drug Administration (FDA), gaining, for example, the creation of the "compassionate use program" and "large-sample trials"—large clinical trials of doubtful scientific value whose costs, borne entirely by industry, often were out of all proportion to the findings that these half-baked studies were able to achieve.

Ironically, one result of this "people pressure" was that many firms reduced their AIDS research. Some worked only on developing substances that had already been discovered, while other firms, such as Lilly and Upjohn, withdrew completely from the field. The AIDS lobby ultimately understood what was going on, and they soon altered their course. Many interest groups possess very knowledgeable experts on the disease triggered by the AIDS virus. Furthermore, they are often excellently informed about technical, statistical, and regulatory aspects of clinical studies and about other key issues in the development of pharmaceutical products. In place of aggressive slogans they set increasingly clear and compellingly formulated arguments.

The relationship between the pharmaceutical industry and the lobbyists on the AIDS scene has improved as a result of this tactical shift. In particular, firms that met the aggressiveness of the AIDS groups with silence and hostility have in the meantime learned to relate in friendlier and more constructive ways with the representatives of AIDS patients, whom their efforts may at long last benefit. But this has not altered the industry's fundamental philosophy. Enthusiasm for AIDS research has diminished even further, not because the technological or scientific conditions for the discovery of new active principles are unfavorable, and not because AIDS is too "small" a disease, but because of the fear of loss of autonomy in the development of new treatments.

Meanwhile, there have appeared other interest groups, though less vociferous, who have taken up the cudgels for those suffering from Alzheimer's disease or from breast cancer. They, too, demand that drugs that are still in clinical trials and about which, therefore, relatively little is known be made widely available to patients. And herein, too, lies a problem for the industry, that such programs will send the already high development costs even higher without a commensurate gain in knowledge. Indeed, a too early, poorly controlled or uncontrolled deployment of medicines holds the great danger that side effects may appear that during the period of testing are still poorly understood and unpredictable, and that such side effects will endanger the development of a product in which millions of dollars may have been invested. A systematic, step-by-step process offers considerably better chances that unwanted outcomes can be analyzed, explained, and avoided.

## Drug Research and Development Outside the Industry?

Yet the increasing number of patient initiatives and their growing ability to articulate their agendas publicly and politically are phenomena that need to be taken seriously. They are expressions of dissatisfaction with the existing health-care system; indeed, they are often an expression of hopelessness and despair. The pharmaceutical industry is but a part of this system, though an important one, to be sure, in the provision of new modes of treatment. If the pressure brought by interest groups increases in certain directions, governments could find themselves forced to react more strongly than has been the case thus far. This could mean that governmental research institutions, such as the National Institutes of Health (NIH) in the USA, the Medical Research Council in England, large, state-financed research institutions in Germany, to name just a few, might initiate their own programs for finding new drugs. Clinical research could as well be carried out under the aegis of governmental health authorities as by the pharmaceutical industry.

Nor should we forget that today there is a competitive industry capable of covering practically the entire field of drug development. Contract research organizations (CROs), research firms that contract with outside organizations to perform a particular task, did a worldwide business in

1994 of $3 billion, with 40,000 employees, a net profit of $300 million, and a capitalization of $4.2 billion, a number that is steadily increasing. There are today over a thousand such firms, most of them in the USA and Europe, but some in Asia as well.[12]

Thus there can be no doubt that drugs could be discovered and developed outside the pharmaceutical industry. The difficulties would be considerable at first, for in mastering the complex and difficult task of producing drugs it is an enormous advantage to have under one roof the functions of research, development, production, quality control, and all the subsidiary functions belonging to these broad categories. However, even outside the pharmaceutical industry there are plenty of opportunities for discovering and developing drugs, and the structures responsible for this situation will become stronger, not weaker.

An industry that becomes disconnected from its true purpose will gradually become replaceable. In place of the large firms, at least in the long run, we might see the birth of small, flexible, discovery-oriented firms. In the USA, in the medical area alone, there are already well over one thousand such firms. Governmentally financed institutes could also undertake the discovery of new drugs. For clinical and preclinical development there is already at hand an industry that depends on contracts. We perhaps cannot envision the provision of modern pharmaceuticals without a pharmaceutical industry. But the dinosaurs also seemed quite comfortably established on the planet; yet they disappeared, and rather suddenly. There will continue to be pharmaceutical research and development as long as people seek to improve their lot in life through medicine, that is, through chemical or gene-therapeutic intervention. But whether this research and development will always take place in the context of today's powerful pharmaceutical industry is by no means certain.

Digitalis purpurea L.

**Figure 2.1.** Foxglove (*Digitalis purpurea*) has been known since the Middle Ages as a medicinal plant. Decoctions of digitalis roots and leaves were used to treat dropsy. (Source: Archiven von PharmaInformation, Basel.)

# 2
∿∿∿

# The History of the Pharmaceutical Industry: Natural Substances as Drugs

To understand the current state of the pharmaceutical industry, we must place the industry in a historical perspective. Most importantly, we should understand the technological and scientific influences that made the development of the pharmaceutical industry possible, influences whose effects are felt even to this day.

## Development of the Pharmaceutical Industry

The pharmaceutical industry arose from two sources. One of these is the aniline dye industry, which evolved in the wake of the industrial manufacture of town gas—gas derived from coal for commercial and residential use. The other is the apothecary, which until well into the nineteenth century produced all drugs prescribed by physicians.

Let us first look at the industrial milieu. Around the middle of the nineteenth century, gas became an important source of energy for industry and trade. It was used more and more, particularly in large cities, for street lighting, and in the home as well it played an important role in lighting, heating, and cooking, a role that grew quickly, becoming the dominant domestic source of energy until well into the twentieth century.

### Town Gas and Coal Tar

This municipal gas was at first obtained almost exclusively from the carbonization of hard coal. In this process coal was heated in the absence of air to about 1000°C. The product of this carbonization process was coke, which was used for heating blast furnaces. But there was a byproduct: coal tar—a black, viscous mass that was at first not much cherished. When it was discovered that this tar contained many substances that could be used as dyes in the textile industry, this waste product suddenly acquired an important economic significance. The interest in tar as a raw material grew even more when it was found that dyes could serve as an important basis for the manufacture of medicines, and that this unsightly mass contained ring and chain carbon compounds that could serve as material for the rapidly developing field of synthetic chemistry. Thus this considerable residue from the manufacture of gas, at first considered a problem, suddenly became important as a raw material, as the source of dyes, drugs, disinfectants, solvents, aromatics, and explosives. As the connection between dyes, which bind to certain fabrics and cells, and chemotherapy became clear (see the section on the birth of chemotherapy in Chapter 3), the aniline dye industry became important as a source of medicines and their precursors. And it is certainly not by chance that the chemical dye factories later branched out into pharmaceutical activities, for which they often became well known. Bayer and Hoechst are examples in Germany of this relationship. In the 1880s and 1890s, Ciba, in Switzerland, began manufacturing pharmaceuticals from components of coal tar. Geigy began in Basel as a drug merchant but developed into an aniline dye concern, thereby broadening its basis as a drug manufacturer and research firm.

## Medicinal Plants and Apothecaries

We have been following one of the strands of the history of the development of the pharmaceutical industry. The other begins with the apothecaries. Until well into the nineteenth century, the apothecary was the institution that filled the prescriptions written by physicians, containing instructions for producing a drug and how it was to be used. With the rise of chemistry in the nineteenth century, the active ingredients of previously known drugs, plants, and plant extracts were prescribed more frequently than the traditional drugs themselves. The first successful experiments leading to the isolation of morphine from opium, emetine from ipecac root, strychnine from the nux vomica tree, caffeine from the coffee plant, and quinine from cinchona bark were carried out by apothecaries in France and Germany. With the acquisition of pure extracts and the characterization of these and other active substances, a new question arose: In order to make these "new" substances available to a wide public, large quantities would have to be produced, and controls would have to be instituted to ensure consistent quality of these products. How was this to happen? Carrying out the necessary chemical processes on a large scale and the establishment and institutionalization of standards—these tasks exceeded the capacities of the apothecaries of the period. A broader outlook was required. To avoid undue competition and to assure themselves a good income, the apothecaries in many parts of Europe had deliberately kept their numbers small. For some populations, particularly rural ones, this resulted in a scarcity of medicines, and often such rural residents had to travel for many hours to have a prescription filled (Figure 1.1).[1]

This new situation suddenly made possible the supply of pure and standardized substances and also, in view of a growing feeling for social justice, encouraged the application of these new possibilities to the population at large. A few apothecaries accepted this challenge and developed themselves into "industrial" apothecaries or pharmaceutical firms. The Golden Apothecary in Darmstadt under Heinrich Immanuel Merck took this route, and today's E. Merck and the American firm Merck Sharp & Dohme can look back on the Golden Apothecary as their precursor. In Berlin the Green Apothecary in Wedding became Schering AG. Carl Engelhardt, in Frankfurt, founded a company from which the two Boehringer firms in Ingelheim and Mannheim arose. A similar story can be told about the Swiss firms of Wander, in Bern, Siegfried, in Zofingen, and Sauter, in Geneva.[2]

Alongside these developments, which were a reaction to a new technological and scientific dynamic, the traditional drug trade continued, growing into the pharmacies that became associated with the newly forming pharmaceutical industry toward the end of the nineteenth century. The notion of trade on an international scale, which later became typical of the pharmaceutical industry, perhaps had its origins here.

Thus there were several factors that came together and interacted in the establishment of the pharmaceutical industry: Technological and scientific impulses led to new industrial processes, which developed new source materials for chemistry and thereby had an effect on the scientific dynamic. Developments in chemistry led in the medical realm to the "scientificization" of therapeutic measures, which hitherto had relied completely on empiricism and tradition. With the reduction of long-known medicinal effects to their individual components there arose the demand to provide such substances, of a consistent quality, to a broad segment of the population. In Europe, many of the pharmaceutical firms were established during a period of economic and political upheaval. The formation of new industries resulted in the movement of large numbers of workers to the cities. The difference in the standard of living between the employers and the employed became an important political theme in the second half of the nineteenth century, leading to the establishment of new laws that insured workers against such vicissitudes of life as sickness and disability, and in which the concept of a work-free old age was realized: In 1883 health insurance was introduced in Germany, in 1884 accident insurance, and in 1889 retirement insurance.

In parallel with the possibility of providing a greater segment of society with physicians and medicines, the state, in a dialogue with representatives of the work force and capitalists, took on the responsibility of actually making this happen. In a broader sense, the foundation of many pharmaceutical firms may have been the entrepreneurial reaction to this new sociopolitical configuration.

## Pharmaceutical Beginnings in America

What was described as the formative stage of modern pharmaceutical companies in Europe applies, of course, also to the United States. Many of the firms that started to operate in continental Europe during the nineteenth century soon expanded their activities to other parts of the world. In this context, the United States became the prime geographical target of many European companies. Hoffmann–La Roche came in 1905 and established its first office in Maiden Lane, in lower Manhattan. In 1929 the first buildings of the Roche campus in Nutley, New Jersey, were completed and the company had thus found its long-term North American home. Sandoz arrived in New York only a few years later and established its campus in East Hanover, in New Jersey, in 1950. Ciba made its entry into the United States in 1921 by purchasing Aniline Dyes and Chemicals; the pharmaceutical division in the United States, which became well known, was established in 1937 and was located in Summit, New Jersey.

Merck, of course, was the most prominent company to come to the United States. A chemist from Darmstadt, where E. Merck had been a firmly established and highly recognized enterprise since 1827, came to New York in 1887 to represent Merck's interest in the New World. His name was Theodor Weicker. Four years after his arrival he was joined by the 20-year-old Georg Merck, grandson of Heinrich Emanuel. In 1899 they acquired land in Rahway, New Jersey, which later became Merck's first American headquarters.

Schering A.G., in Berlin, started to sell diphtheria antitoxin in the United States through their New York agents, Schering and Glatz. Both Schering and Merck were seized by the "Alien Property Custodian" during World War I and subsequently became private American companies, totally independent of their German progenitors.

Thus a significant portion of the pharmaceutical industry in the United States has European roots; that is, they originated from pharmacies or from dye companies.

The majority of companies that form today's American pharmaceutical industry, however, were born on the American continent: SmithKline and French; Wyeth; E. Squibb and Sons; Parke, Davis and Company; Eli Lily; G.D. Searle; Lederle Laboratories; the Upjohn Company; and Abbot Laboratories—to name only some of the most important ones.

As we will see, these companies originated in somewhat unorthodox ways. Unlike the European examples mentioned above, the beginnings of American companies did not so clearly relate either to dye divisions of chemical firms or to pharmacies. Very often, the founders of these companies were trained pharmacists, like Eli Lilly. The driving force for the formation of Eli Lilly and Company, however, was not so much the need for standardized drugs or the ability of a well-established pharmacy to provide them. Not a pharmacy going industrial, but a pharmacist starting his own business by selling and to some extent making the pills, fluid extracts, elixirs, and cordials of the day—that was in a nutshell the situation that led to the formation of many American companies, Eli Lilly; SmithKline and French; and Parke, Davis among them.

Even in places that had built a pharmaceutical tradition in the European sense, this pattern prevailed. Early in American history the city of Philadelphia became what some have called the "cradle of American pharmacy." Christopher Marshall, an Irishman who became a Quaker, established the first apothecary shop in Philadelphia. It carried the "Sign of a Golden Ball" as its trademark. During the first half of the nineteenth century, Europeans brought certain pharmaceutical skills to Philadelphia. John Farrar, for instance, from England, and a Swiss, Abraham Künzi, began to make quinine in 1818. Around the middle of the century, two people with similar names, one Seitler and one Zeitler, the first a Frenchman and the second German, opened a drug business in Philadelphia. Since Seitler spoke only French and Zeitler only German, they needed an interpreter. Their accountant, George Rosengarten, played this role and later bought them out. George K. Smith, together with his bookkeeper,

Mahlon N. Kline, formed a drug firm that had nothing to offer except trading with contemporary medicines of doubtful quality. Nevertheless, this company, which later merged with a perfume manufacturer, Harry B. French, became the famous SmithKline and French, which after their merger with Beecham of England formed SmithKline and Beecham, one of today's most prominent pharmaceutical companies. The Wyeth brothers established their company, which later was to become part of American Home, in 1860. The need to produce penicillin as part of the American war effort gave Wyeth an unexpected chance at prosperity. At a time when penicillin had to be produced from molds growing on surfaces, Wyeth scientists proposed to use mushroom cellars that had been set up by gourmets near Philadelphia. For a short period, Wyeth became the world's leading producer of penicillin by a procedure called the "milk bottle method." Afterward, fermentation in suspension cultures became the basis for industrial production of this and other antibiotics.

Without any doubt, E.R. Squibb has from its beginnings been one of the most quality-oriented companies of American descent. In the first place, this was due to the personality of the founder, Dr. Edward Robinson Squibb, a physician. Squibb was the son of Quakers whose ancestors had come to this country with William Penn and were charter members of the first Society of Friends in this country. Squibb held himself and his collaborators to very strict ethical standards. He was the forerunner of medical marketing practices based on truthfulness and completeness of information. He attended medical meetings, visited doctors' offices, and even published a magazine for physicians. Ether for surgical use was the company's first product. Later, E.R. Squibb became well known for its work on rauwolfia drugs, as well as on insulin and vitamin preparations.[3]

There are many similarities between the beginnings of European and American pharmaceutical companies. The unique feature that distinguishes most American companies from their European counterparts seems to be the prevalence of entrepreneurial impulses over scientific and strategic motivations. In Europe most, if not all, companies, arose as logical diversifications of the dye divisions of chemical companies or as continuations of apothecaries that responded to growing needs in an organized fashion. There always appeared to be a scientific or technical opportunity for these strategic mechanisms. Commercial opportunity was the prime force that led to the foundation of most American drug companies. There was little scientific or technical logic or ambition in these beginnings. Compared with their European counterparts, American companies were rather crude creatures, which during the first stages of their life had little or no connections to academia or to the scientific community at large. In America these links did not develop until the time between the two World Wars. When German firms interrupted drug supplies to the United States in World War I, the deficiencies of the American pharmaceutical industry became very obvious. At this time, American companies started to turn seriously to drug development and later to research. Consequently, the bias of basic scientists in the United States against the "pill-peddlers" started to erode and to give way to a more realistic and constructive relationship. Also, industry leaders recognized that they would be able to sustain and grow their companies only on the basis of new drugs, which they would have to provide themselves through research and development. The Second World War with its existential pressures brought academia and the pharmaceutical industry even closer together, and by the end of the war, the American industry was on its way to taking the lead in drug discovery and development.

Today, the great American companies and their European peers can be regarded as equals. Global competition, scientific opportunities, and political pressures affect them all more or less in the same way.[4]

## Drug Research and Drug Therapy in a Time of Scientific and Technological Change

There are two aspects to drug research: On the one hand there is the scientific, which is to say the theoretically and methodologically assured, way to the production of chemical compounds. Such compounds can be naturally occurring substances, that is, existing in living organisms, or they can be synthetically produced. After more than a hundred years of experience in both directions, we know today that the natural compounds are often much more effective and better tolerated than their synthetic counterparts: The semisynthetic antibiotics, here above all β-lactams, but also tetracyclines and macrolides, are good examples of the fruitful complementation of two chemical traditions.

The second aspect of drug research is testing to determine whether the chemical compounds produced are effective. The measurement, classification, and testing of the effectiveness of drugs is the concern of pharmacology. Pharmacology began its life as an experimental discipline with a sharp focus on physiology, that is, on a quantitative understanding of bodily functions. Today, pharmacology is more focused on disease; it is no longer content to measure the effects of chemical compounds on normally functioning bodies, for which healthy laboratory animals are the ideal model, but it is also interested in the possibility of correcting malfunctioning physical processes by pharmacological means. For this it requires animal models, by which is understood the possibility of experimentally creating particular disease symptoms in animals. Animal models should imitate in part human diseases, and they should be "robust," that is, entirely reproducible. Furthermore, today's pharmacology has outgrown merely describing the effects of drugs on animals. It has developed in clinical pharmacology a complete apparatus for measuring the effects of drugs on human beings. In parallel to the development of experimental pharmacology, which began with the analysis of normal function and then progressed to the study of models of disease, clinical pharmacology at first confined itself to the study of normal functioning in healthy subjects. Only later did it carry over its experience to models of human disease.

Experimental pharmacology developed its methods using a very narrow spectrum of substances that had achieved recognition as traditional drugs. Drug research as it is known today began at the moment when two mature scientific disciplines that were ready to engage in dialogue—chemistry and pharmacology—entered into a relationship governed by scientific principles and dedicated to bringing forth new medicines. For this meeting an institutional framework was required. Such a framework did not exist at the universities at the end of the nineteenth century. Moreover, no one affiliated with an academic institution would have been willing and able to create one. It was the emerging pharmaceutical industry that provided the requisite milieu. One could argue that the joining of forces of chemistry and pharmacology in the search for new drugs was one of the important motives for the founding of pharmaceutical firms.

The fact that the European, and most of all the German, universities took so long to recognize and develop interdisciplinary fields of study is also illustrated by the fact that they have only recently (as of 1997)—that is, a good century after the establishment of most of the well-known pharmaceutical firms—begun to integrate traditional pharmaceutical sciences and pharmacology in the wider framework of drug research, a field that also includes aspects of chemistry, genetics, and biochemistry.[5]

## Natural Substances as Drugs

The use of plants or plant extracts for treating human illness seems to be almost as old as mankind itself. In any case, there is evidence of the use of medicinal plants among the earliest traces of human culture. Thus, for example, a Chinese emperor, Shen Nong, who appears to have lived about five thousand years ago, left to posterity the descriptions of 365 medicinal plants. This early collection already contained plants similar to those of the genus *Ephedra*, which were used in the treatment of bronchial diseases, as well as *Rizinus communis* and the opium poppy, *Papaver somniferum*. These plants and many others were discovered and described as drugs independently in widely separated parts of the world. Many are attested in Assyrian, Egyptian, Greek, and Central European sources. Perhaps the most comprehensive and informative early document concerning drugs and other medicinal substances in ancient Egypt is the Ebers papyrus. This scroll, a foot wide and almost sixty-five feet long, was found between the knees of a mummy in a grave in Thebes and was brought to Europe by the archeologist Georg Ebers. Written about 3500 years ago, it contains over eight hundred sets of instructions for treatment, many of which recommend the use of plants, others the use of animal organs or minerals. Incantations and healing songs are also part of the therapeutic regime—an early example of "holistic therapy."[6]

We may conclude from this that empirical knowledge of the medicinal use of plants and other organisms (and minerals as well) has been sought and described in all eras. Some of these descriptions have been passed down to the present day, where they have established a place for themselves in a scientifically based system of drug research and medical practice. At the close of the eighteenth and throughout the nineteenth century many traditional medicinal plants were subjected to chemical analysis. A complete description of this work, which at first took place primarily in France but then spread to England and later to Germany, would be beyond the scope of this brief introduction. Therefore, we shall here confine ourselves to showing, with a few examples still relevant today, how a system based on empirical knowledge evolved into another one in which the active principles could be categorized by their chemical structure and later by their pharmacological properties (see Table 2.1).

## Papaver Somniferum: Opium

Opium, the juice of the opium poppy *Papaver somniferum*, is one of mankind's oldest effective drugs. We have already mentioned that the calming effects of the poppy were familiar to the Chinese emperor Shen Nong. Poppy was also mentioned in the Ebers papyrus. In Central Europe the remains of poppy plants have been found in Stone Age settlements. This

Table 2.1.  Natural substances and their derivatives in medical practice.

| | 1850 | 1860 | 1870 | 1880 | 1890 | 1900 | 1910 | 1920 | 1930 | 1940 | **1950** | 1960 | 1970 | 1980 | 1990 | **2000** |
|---|---|---|---|---|---|---|---|---|---|---|---|---|---|---|---|---|
| *Empirical medicine* | Quinine, Opiates/morphine digitalis | | | | Ephedrine | Atropine and other alkaloids | | | | | Ergotamine | Vinca-alkaloids[1] | | | | |
| | | | | | | Digoxin | | | | | | | | | | |
| *Extracts from plants, animals, and minerals* | | | Salicylic acid | | | ASS | | | | | | | | | | |
| *Scientific medicine on the basis of reproducible results of biology, chemistry, physics, and microbiology/biotechnology* | | | | | | Animal antibodies (sera) | | Insulin, other peptide hormones | | | Heparin, streptokinase[2] | | Monoconal antibodies[3] | | | |
| | | | | | | | | | Vitamins | | | | Retinoids[4] | | | |
| | | | | | | | | | | | Cortisone, sex hormones[5] | | | | | |
| | | | | | | | | | | | | Penicillins, cephalosporines, tetracyclines[6] Streptomycin, rifampicin | | | | |
| | | | | | | | | | | | | | | Mycins, Rubicins, Paclitaxel[7] | | |
| | | | | | | | | | | | | | | Cyclosporin A | Mycophenolate[8] | |
| | | | | | | | | | | | | | | | Acarbose Lipstatin[9] | |

[1] Alkaloids
[2] Pain relievers/anticoagulants
[3] Peptides
[4] Vitamins/Vitamin A derivatives
[5] Sex hormones, steroid hormones
[6] Antibiotics, anti-infectives
[7] Antitumor substances
[8] Immunosuppressives
[9] Metabolic regulators

*The first letter of the name of the substance is positioned at the year of introduction.

plant, already mentioned by Homer (ninth century B.C.E.), was given its first comprehensive description by Theophrastus (371?–287? B.C.E.) in his *Historia Plantarum*. From a later period, in ancient Rome, come instructions on the best method of extracting the juice of the poppy. The calming and anesthetic effects of poppy thus represent an ancient knowledge, accumulated and confirmed apparently in different places and at different times. Wilhelm Hufeland, a representative of romantic medicine, named opium "a great, secret, extraordinary agent, even now incomprehensible in its effects." Other important physicians of the modern era, from Thomas Sydenham to Gerard van Swieten, have expressed themselves at length—though often contradicting one another—on the properties of this drug.

The history of what we might call modern research on opium began only in 1803, in Paris and in Paderborn, in both cases in apothecaries. Jean François Derosne was the owner of a Parisian apothecary. In a letter to the *Société de Pharmacie* he reported that he had isolated a crystalline salt in the course of developing a new analytical test for determining the presence of opium. Derosne could not say much about the nature of his crystals. He believed that he could only exclude the possibility that the new component was a plant acid. One year later, another Frenchman, Armand Seguin, reported on the isolation of a vegetable acid and a narcotic in crystalline form obtained from opium sap. Seguin's work did not attract much attention at first, and it was not published until 1814, in *Annales de Chimie*. By this time reports by other researchers about similar discoveries had already appeared.[7]

In the same year, Friedrich Wilhelm Sertürner, a young Austrian working in Westphalia, completed his pharmaceutical apprenticeship at the Paderborn Court Apothecary. For the next two and one-half years Sertürner worked in the Eagle Apothecary, in Paderborn, where he studied the composition of opium (Figure 2.2). At almost the same time as Seguin, Sertürner, too, found an organic acid that had no narcotic effect, which he named meconic acid. After the alkalization of the liquid from which he had precipitated the meconic acid, Sertürner obtained yet another substance, which crystallized in alcohol. Unlike meconic acid, this second component was shown to have a narcotic effect on dogs. Sertürner named the substance *principium somniferum,* that is to say, the soporific principle, and in a paper published in *Johann Trommsdorffs Journal der Pharmazie* indicated that it was of alkaline nature, in that the new substance neutralized free acids. The alkaline nature of the substance was verified in further studies. Only in 1815 did Sertürner begin a chemical analysis of the compound.

Figure 2.2. Friedrich Wilhelm Adam Sertürner (1783–1841) was the discoverer of morphine. (Source: *Die Apotheke. Historische Streiflichter.* 1996, Editiones Roche. F. Hoffmann–La Roche AG, Basel, p. 247. Universitätsbibliothek Basel, Portraitsammlung.)

His results, published in 1817 in the respected journal *Gilberts Annalen der Physik,* showed the presence of carbon, hydrogen, oxygen, and nitrogen. Unlike his earlier articles, this one was met with great interest. No one less than Joseph Louis Gay-Lussac had the article immediately translated into

French and arranged for its publication in the *Annales de Chimie*. He himself wrote an editorial in the same issue of the *Annales* in which he pointed out the—at the time surprising—alkaline reaction of the narcotic substance and simultaneously postulated that more alkaline substances with bioactive properties would be found in plants: Many of the existing substances obtained from plants in their unpurified form contained nitrogen and exhibited an alkaline reaction. Gay-Lussac suggested that the suffix *-ine* be appended to such vegetable bases and that the term *morphium* suggested by Sertürner be changed to *morphine*. This suggestion was accepted, as was the proposal made in 1818 by Wilhelm Meissner, in Jena, that such alkaline substances derived from plants be called *alkaloids*.

In the following years, France became the center of alkaloid research. Gay-Lussac asked his colleague Jean Robiguet, who worked at the *Ecole Supérieure de Pharmacie,* to reproduce Sertürner's experiments. Robiguet found differences in the reaction of the salts discovered by Derosne and Sertürner, and from this he concluded that there must be more than one plant alkaloid involved. This conjecture turned out to be correct. The compound isolated by Derosne was first named *narcotine* and later *noscapine*. It has no narcotic effect, but it is effective as a cough suppressant, for which purpose it is still used.

Sertürner himself had already known that morphine, or morphium as he called it, comprised only half of the alkaloids in *Papaver somniferum* and that the alkaloid portion of opium juice was only about ten to twenty percent of the total. In 1848, G. Merck isolated another alkaloid from opium, one that produced sleep in animals and that later established itself as an antispasmodic. Its effect of suppressing contractions of the smooth muscles of the gastrointestinal tract was discovered in 1917 by David Macht at the Johns Hopkins University. Papaverine, as the substance was named, became the chemical precursor for many effective drugs that suppress spasmodic and/or other brain triggered involuntary movements. An example is Akineton®, a substance that suppresses involuntary extrapyramidal movements, such as appear after treatment with phenothiazines.

The correct chemical structure of morphine was not given until 1923, by John Gulland and Robert Robinson (Figure 2.3). Nonetheless, researchers had already begun to search for better, more effective, derivatives of morphine by varying the functional groups of the molecule. Naturally occurring opium alkaloids such as codeine, the methyl ester of morphine, pointed the way to the synthesis of other esters, for example the ethyl ester, which was long in use as a cough suppressant (antitussive).

**Morphine**
Opioid analgesic

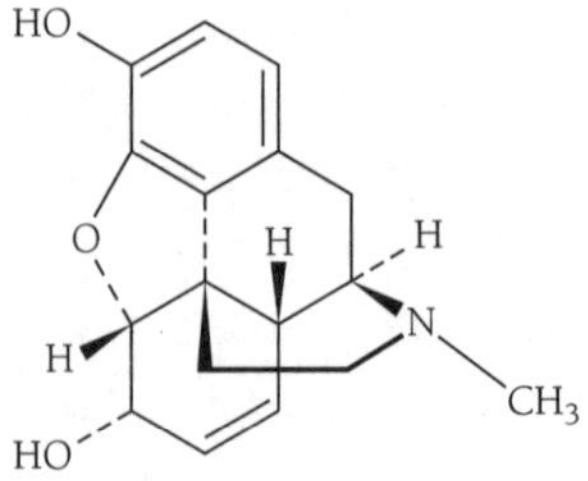

Figure 2.3.

Diacetylmorphine, or heroin, was at first known as a pain reliever (analgesic) with a weakened depressive effect on breathing before its dangers as an addictive poison became known (Figure 2.4). Rudolf Grewe, of Hoffmann-La-Roche, was the first to succeed in synthesizing the complete morphine structure, which was known as N-morphinan. The 3-OH derivative of N-morphinanmethylester, levorphanol, became under the name "Dromoran" a strong analgesic that could be administered orally. The dextrorotatory isomer of this compound, dextromethorphan, has no analgesic effect, but it is still used as an antitussive. Recently, dextromethorphan and dextromorphan have aroused interest as potential "neuroprotective" agents. These or similar compounds can help to reduce the extent of cerebral damage after trauma, poisoning, or oxygen deprivation. The interest in morphine derivatives lasted a long time, especially in the German and Swiss pharmaceutical industries. The hopes of developing analgesics based on the structure of morphine without negative side effects on breathing were

**Diacetylmorphine, heroin**
Cough-suppressant substance
Opioid analgesic

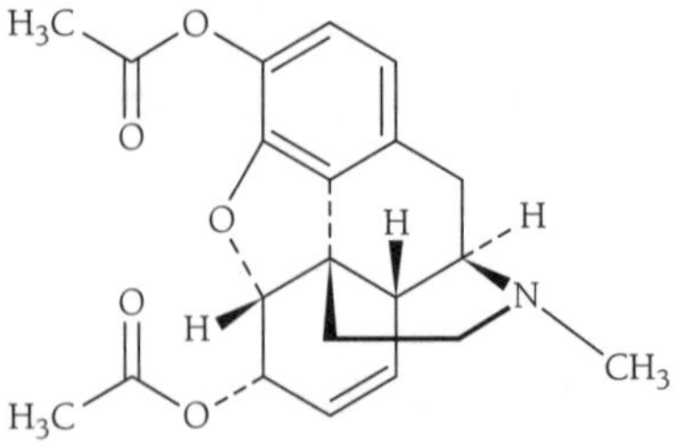

Figure 2.4.

not met with adequate success. However, it is worth noting that his involvement with new pethidine and methadone derivatives led Paul Janssen in 1958 to the discovery of a completely different effect: Haloperidol (butyrophenone) has proven itself a highly effective antipsychotic.

A precise knowledge of the structure of morphine was an important, but by no means indispensable, prerequisite for working on this molecule. The formulation of morphine preparations of standardized purity was, however, a decisive prerequisite for researching opium alkaloids in a methodical dialogue, beginning in the middle of the nineteenth century, among chemistry, pharmacology, and clinical research.

But it is equally uncontested that the precise knowledge of the structure of morphine and the development of a way of synthesizing the basic morphine structure opened the modern era of semisynthetic and synthetic derivatives of opium alkaloids, including the antagonists and partial antagonists.

From these examples of opium, the discovery of morphine, the systematic research into the effects of morphine, the analysis of the structure of morphine, and the consequent bifurcation of directions of research in chemistry (and pharmacology), one can recognize how deeply we are connected today, in the era of chemical and biomolecular research, to experiences and observations that go back hundreds, and indeed thousands, of years.

## Quinine

The view voiced on the occasion of the publications of Sertürner and Gay-Lussac that there remained many plant alkaloids to discover spurred Joseph Caventou and Joseph Pelletier, the former still a chemistry student of Gay-Lussac, the latter already a professor, to search for more such substances. The isolation of emetine from ipecac root was the first fruit of these efforts. Ipecacuanha had been known since the seventeenth century as an emetic and as a remedy for diarrhea and dysentery. This knowledge originated among the inhabitants of Peru and Brazil. The isolation of strychnine from a variety of species of the genus *Strychnos* is also due to Pelletier and Caventou. They confirmed the hypothesis expressed by Linnaeus that plants belonging to the same family would be found to have similar pharmacological properties. Their most memorable and significant achievement, however, was the isolation of quinine from the bark of trees of the genus *Cinchona*. This bark had been imported into Europe at the beginning of the seventeenth century, two centuries before the isolation of quinine (Figure 2.5). An Augustinian monk had reported at the time that the native population used this bark for treating malaria. This "Jesuit's bark," preparations of which were evil-tasting and had many side effects, primarily in the stomach and intestines, became popular as the result of a report by an English apothecary's apprentice, Robert Talbot. Later (1712), the therapeutic properties of this bark were described in detail by Francesco Torti, in Modena. Many varieties of fever could not at that time be differentiated by their causes, but rather by the courses they took. Thus one differentiated between "continuous" and "periodic" courses of fever and remarked that the latter type was frequent among individuals who visited warm climates. Among the periodically appearing forms of fever one distinguished between the less serious "intermittent" and the more serious "recurrent" courses. As John Brown, a collaborator of William Cullen, the well-known, at that time, founder of the Glasgow Medical

ÆGROTAT LIMÆ CONIUX CHINCONIA FEBRIM
CORTICE MIRANDO POCULA TINCTA FUGANT

**Figure 2.5.** In the Ospedale di Santo Spirito, in Rome, there is a series of frescoes depicting the introduction of quinine in the fight against malaria. Supposedly, the Countess of Cinchon, wife of the viceroy of Peru, brought quinine to Europe. The various types of cinchona trees, the source of the alkaloid quinine, are named for her. (Source: *Die Apotheke. Historische Streiflichter.* 1996, Editiones Roche. F. Hoffmann–La Roche AG, Basel, p. 214. Universitätsbibliothek Basel, Portraitsammlung.)

School, postulated, these periodic forms of fever, especially of the intermittent type, appeared frequently among sickly, or "asthenic," patients. Cinchona bark was therefore classified as a tonic for the nervous system, and its effect against malaria was initially attributed to a stimulant effect on the brain. In a paper published in 1820 Pelletier and Caventou reported on their isolation of two alkaloids from gray and yellow cinchona bark: cinchonine and quinine. At that time Pelletier and Caventou committed themselves decisively to studying the pure cinchona alkaloids instead of what had previously been the usual mixture, and they thereby laid an important cornerstone for the development of pharmacology.[8] That this thought remains current to this day is reflected in the proverbial saying from the pioneer days of molecular biology: Don't waste pure thinking on impure substances.

Immediately on the heels of the preparation of pure quinine there followed a proof of its effectiveness against malaria (Figure 2.6). The pure alkaloid had an advantage over the bark extract not only in its effectiveness, but also in its tolerability. It was soon held in high regard and was manufactured and used in large quantities. Soon there was a demand for quinine on the order of several tons per year. However, it was required to produce the substance in a uniform and reliable quality, and this could only be accomplished on an industrial scale. The provision of quinine, like the preparation of morphine of uniform quality, became a significant challenge for the pharmaceutical industry that was emerging at the end of the nineteenth century.[9]

Quinine, like morphine, turned out to be extremely fruitful for pharmaceutical research. Its reputation as an antipyretic, that is, a fever-reducing

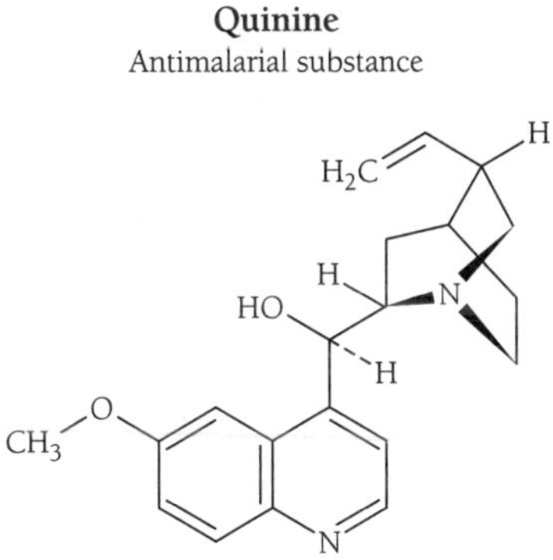

**Quinine**
Antimalarial substance

Figure 2.6.

remedy, came from times when the root causes of malaria were unknown. Of course, at that time nothing was known about quinine's effectiveness against the plasmodia protozoa. Thus the attempt to use quinine as a starting point in the search for new fever-reducing remedies rested on a misunderstanding. Nonetheless, it was successful! By studying the decomposition products of quinine, the quinolines, Ludwig Knorr eventually succeeded, by way of semisynthetic quinine derivatives, in the more-or-less accidental synthesis of phenazone, or *antipyrine*, and thereby to the most-used medicine of the turn of the century.[10]

The results at the beginning of the twentieth century of self-experimentation, which consisted in taking quinine, by a patient with atrial fibrillation convinced the prominent Dutch cardiologist Karl Wenckebach of the antiarrhythmic effect of quinine. It later turned out that quinidine, an isomer of quinine, was an even more effective and better-tolerated antiarrhythmic, that is, a medicine for restoring normal cardiac rhythm.[11]

Furthermore, aminoquinoline, a decomposition product of quinine obtained by the substitution of two aromatic rings, became an important source for producing new agents against malaria that unlike quinine, which inhibits the reproduction of pathogens in red blood cells, the erythrocyte phase, reduces the survivability of parasites in the liver. A combination of both substances leads in many cases to a complete cure from malarial infection. Pamaquine was the first quinoline derivative. Later came atebrine (an aminoacidin that targeted both erythrocytes and liver cells) and other quinoline derivatives.

## Glycosides Acting on the Heart

Like the plant alkaloids, glycoside drugs have a long history. The longest known seems to be hellebore (*Helleborus niger*). It is even mentioned in Greek mythology. Paracelsus described hellebore as a stimulant of mucous membranes and a strong laxative. He also mentioned a dehydrating, diuretic action.[12] The sea onion (*Scilla maritima*) seems to have long been used as a diuretic. Pythagoras mentions it, and his knowledge of medicinal plants presumably came from Egypt.

On the other hand, the medicinal use of lily of the valley (*Convallaria majalis*) appears to be a discovery of the modern era. Allusions to a "heart-strengthening" effect can be found in the 1530 *Herbarum Vivae Eicones* of Otto Braunfels and a little later (1542) in Leonhard Fuchs's *Historia Stirpium*.

It is interesting to note that there are no indications of an ancient knowledge of the most effective group of plant glycosides, namely the active substances in foxglove, the digitalis plant. However, in the British Isles foxglove has been known as a medicinal plant since the Middle Ages. It was mentioned in an old Celtic manuscript, the Irish *Meddygon Midway*. The plant was also described in the above-mentioned work of Leonhard Fuchs.[13] In England the drug must have been known as a folk remedy in the seventeenth and eighteenth centuries, since William Withering, the English physician who is considered the discoverer of the medicinal properties of foxglove, himself described how he became acquainted with foxglove as a medicine through a family recipe that was in the possession of an old woman in Shropshire. Withering immediately identified foxglove (*Digitalis purpurea*) among the more than twenty plants mentioned in the recipe as the most likely active component. A decoction of digitalis roots and leaves, that is, an extract obtained by boiling down, proved itself an effective remedy in the treatment of "dropsy," conditions in which fluid accumulates in the tissues, presumably as a result of cardiac incapacity. Withering himself, in a letter of 1777 to a surgeon in Worcester, wrote of a diuretic effect. Accumulation of fluid in the thorax (hydrothorax), in the pleura (pleural effusion), in the pericardium (pericardial effusion), or in peripheral tissues (edema) responded well to the decoction of *Digitalis purpurea* that Withering prescribed. In his reports Withering mostly commented on the diuretic effects, although he also mentioned a direct action on the heart that could be used for medicinal purposes.[14]

Damage to the heart that could lead to cardiac insufficiency must have been more prevalent then than in our time. This is surely because of the numerous diseases caused by bacterial infection, above all infection by streptococci, which were classified as "fevers" and which very often led to infections of the heart muscle and valves, developing into cardiac decompensation. However, Withering and his contemporaries also observed the side effects of digitalis therapy, above all the typical signs of overdosage, or "saturation": nausea, vomiting, diarrhea, dizziness, seeing colors, and slowing of the heart rate (bradycardia). These effects gave digitalis the reputation of being dangerous, although the drug was judged very positively, despite the risks involved, by such physicians as Wilhelm Hufeland.[15]

Richard Bright, of London, was the first to distinguish between accumulation of fluid caused by cardiac insufficiency and that caused by insufficiency of kidney function. In this connection he pointed to proteinuria as a symptom of edema caused by disturbances in kidney function. One should have quickly come to the conclusion that digitalis is suited only for treatment of those cases of fluid accumulation due to cardiac problems, but this did not occur until well into the nineteenth century. In 1855, there was still no mention of this distinction in the *Elements of Materia Medica and Therapeutics*. On the other hand, we find such counsel in the 1890 *Dispensatory of the United States of America*.[16]

In Germany, knowledge of the correct application of digitalis is due most of all to the work of the internist Lukas Schönlein (1793–1864).[17] Ludwig Traube (1818–1876), Schönlein's student and like him a clinician, but also an experimental pathologist, described the pulse-slowing effect of digitalis extract on a dog paralyzed by curare, an effect that could be canceled by severing the vagus nerve.[18] The significance of Traube's findings was strongly influenced by the discoveries of E.H. Weber and C. Ludwig, who had just discovered the cardiac nerves. It was the English pharmacologist J. Milner Fothergill who first freed himself from the idea of a connection between cardiac nerves and muscle contractions and made the hypothesis that the effect of digitalis rested entirely on its ability to strengthen the contractions of the heart.

In order to confirm this hypothesis, new methods were necessary, methods that were first developed at the Oswald Schmiedeberg Institute, in Strasbourg. In a paper that appeared in 1872, Rudolf Böhm[19] described the cardiac effects of digitoxin and digitalin in an isolated frog heart and in situ. Severing the nerves that suppress cardiac function appeared to have no influence (such as lengthening the diastole, strengthening the systole) on this effect. The final proof of a direct effect of digitoxin on the heart muscle was provided in 1880 by Francis Williams, who worked with Schmiedeberg in the Strasbourg Pharmacological Institute. He developed a method that made it possible to measure quantitatively the process of contraction in isolated frogs' hearts, thus eliminating any systemic physiological effects. Williams verified and enlarged the set of observations made by Rudolf Böhm. Later, Heinrich Dreser, in collaboration with Schmiedeberg, refined these techniques by building into the experimental procedure a variable and controllable electrical resistance.[20] Dreser became, after his habilitation in Tübingen and subsequent residencies in Bonn and Göttingen, the first "industrial pharmacologist." In 1897 he was given the opportunity by the dye manufacturer Bayer to set up a pharmacological laboratory. Heinrich Dreser, whom we shall meet again in connection with aspirin, personifies, one might say, one of the links between the field of experimental pharmacology, which was gradually coming into its own, and the concept of pharmaceutical research that was developing within industry.

In 1820 the search had already begun for the active components of the digitalis plant, but it was not until 1841 that E. Homolle and Théodore Quevenne, the chief pharmacists of the Charité, in Paris, produced a still impure, but nonetheless crystalline, material that was already more active than the previously known preparations. Presumably, it was composed principally of digitoxin. The substance was named digitalin. Oswald Schmiedeberg isolated digitoxin in 1875, in Strasbourg. To do this he used techniques that had earlier been used by C.A. Navitelle and with which in 1869 Navitelle had isolated a substance that he designated crystalline digitalin. Although the production of pure digitalin and digitoxin made it easier to assess the therapeutic and experimental effects of these glycosides, it was almost another half century before the structures of these substances were known. In 1928 and 1929 Adolf Windaus and his colleagues published the correct chemical structures of digitoxin and digitalin. Finally, the palette of digitalis glycosides was enlarged by the description of digoxin from *Digitalis lanata* and strophanthin from the African trees *Acokanthera ouabaio* and *Strophanthus gratus*.[21] Both glycosides proved to have therapeutic value. In many places digoxin was soon preferred to digitoxin because it binds less strongly to plasma proteins and other cellular components, is more quickly excreted, and because of these characteristics is less susceptible than digitoxin to being stored in the organism. Aside from their clinical significance, the digitalis glycosides became model substances for the rapidly growing field of experimental pharmacology. We shall have more to say about this later.

## Aspirin

The discovery of aspirin came about by two separate routes: On the one hand, it was a consequence of the search for a fever-reducing drug from willow bark. On the other, it resulted from the search for new, synthetically produced compounds with fever-reducing and pain-killing properties. When these two research paths crossed, acetylsalicylic acid was found, and at this crossroads its remarkable powers were recognized.

The fever-reducing (antipyretic) and anti-inflammatory properties of willow bark had been discovered, forgotten, and rediscovered many times over the millennia. Hippocrates mentions preparations from the bark of

*Salix alba* as analgesic and fever-reducing remedies. In the modern era, willow bark was discovered by an English vicar, who in searching for fever-reducing remedies began with the magical—though in his case we may say Christian—assumption that Providence has placed cures for diseases in the midst of their causes. Edward Stone, which was the devout man's name, had observed that people who lived in the vicinity of damp or swampy areas were particularly susceptible to fever. Since willows grew in swamps, and moreover, there was at this time, the early 1760s, an example of an important medicine obtained from tree bark, namely quinine, Stone decided to employ willow bark against various types of fever. He pulverized a pound of dried willow bark and administered the resulting powder mixed in water, tea, or beer to fifty individuals sick with fever. These "clinical trials" lasted several years. Stone observed that with few exceptions, administering his medicine resulted in lowering the fever and lessening pain. He published his findings in 1763 in a journal of the Royal Society of London. In 1821, German researchers isolated from willow bark a yellow, bitter-tasting crystalline substance that they named "salicin." Today, we know that what they had found was sodium salicylate. Ten years later a group of French chemists succeeded in synthesizing salicylic acid. In 1874, Thomas MacLagan, a Scottish physician, administered salicin to patients with acute rheumatic fever. A short while before, Carl Buss, in Berlin, had carried out similar experiments. Both men saw clinical improvements and were convinced of the therapeutic value of salicin and salicylic acid, respectively, in the treatment of rheumatic fever. Marcellus von Nencki, in Basel, had shown in 1870 that salicin is metabolized in the body to salicylic acid. Buss, at least, knew already that salicylic acid was the active principle of the longer-known salicin.

Medicines for reducing fever and suppressing inflammation made up the most attractive and fastest-growing segment of the young pharmaceutical industry at the end of the nineteenth century. In 1883, Ludwig Knorr had synthesized antipyrine and assigned the rights for this substance to Messrs. Lucius and Brüning, in Hoechst. In 1866, two Alsatian physicians, Drs. Kahn and Hepp, in the search for antiparasitic compounds, came upon acetanilide, a product from coal tar possessing fever-reducing properties, though ineffective against parasites. They assigned the rights for this substance to the firm Kalle, in Wiesbaden, who already held a patent for its production. Acetanilide was produced under the trade name "Antifebrin." The success of Antifebrin encouraged Duisberg, who at the time was responsible for the Bayer laboratories, to search for his own antipyretics. He started from the observation that there is a chemical similarity between acetanilide and p-nitrophenol, a waste product of dye manufacture from coal tar that was available in large quantities. In 1888 the dye manufacturer Bayer produced its first pharmaceutical product, acetophenetidin, or phenacetin. Phenacetin was the first drug to be conceived, synthesized, developed, and marketed by a private firm. Phenacetin was a great medical and commercial success. This success encouraged Duisberg to set up a laboratory in Elberfeld whose sole purpose was to be the discovery of new medicines. This new pharmaceutical department comprised a synthesis laboratory, directed by Arthur Eichengrün, and a pharmacological division under the direction of Heinrich Dreser, whom we have already met as a student of Oswald Schmiedeberg. Eichengrün had the idea to look for a second derivative of salicylic acid, one that had fewer side effects on the stomach and intestines than salicin or pure salicylic acid itself. For this project he took on a young chemist, the twenty-nine-year-old Felix Hoffmann. On October 10, 1897, Hoffmann met with success. There is an entry for that date in a laboratory notebook about the synthesis of acetylsalicylic acid: Aspirin, the most successful medicine of all time, in use for over a hundred years, had been born (Figure 2.7). To be sure, Hoffmann was not the first to synthesize acetylsalicylic acid. It had already been produced in 1853 in crude form by Charles Frédéric Gerhardt, in France, and sixteen years later by Carl Johann Kraut, in Germany. However, its superior properties were recognized only after this compound had been found as the result of a deliberate search for new fever-reducing and pain-killing substances. Aspirin did not come into the world without pain. At first Dreser rejected the substance. Given the pharmacological methods of the time, he could discover no advantages over other antipyretics. At the time, a heart-damaging effect was ascribed to the salicylates, which Dreser believed to have found as well in the case of aspirin. Eichengrün, however,

**Acetylsalicylic acid, Aspirin®**
Analgesic

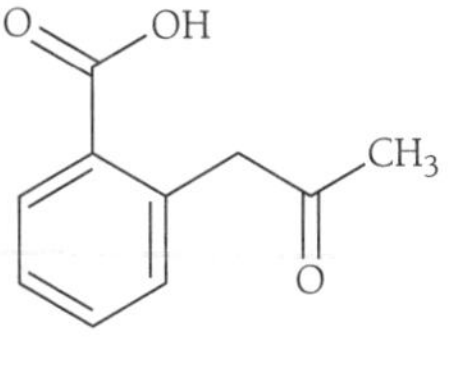

Figure 2.7.

was not satisfied with Dreser's negative judgment. He took the substance himself, determined that if he suffered no ill effects he would then try it on patients. The fact that the Berlin physicians were of greatest help and that one of them, Dr. Felix Goldmann, wrote an enthusiastic report on the basis of his findings was put down by Dreser to "the usual Berlin braggadocio." But at this moment Duisberg, on the basis of a report from Eichengrün, entered the fray and demanded that the substance be tested by an outside pharmacological laboratory. This was done, and the substance passed with flying colors, after which Dreser agreed to further development. Salicylic acid also occurs in a shrub that grows in clusters and produces an abundance of white blossoms in the spring. The shrub is called spiraea, and salicylic acid was therefore called *Spirsäure* (spiraeic acid) in Germany. From *Acetyl-Spirsäure* was derived in 1899 the name *Aspirin*, which in that year set out on its ongoing triumphal march. At the beginning of the twentieth century and well into the middle of the century, aspirin became a sort of cult drug, especially in the United States of America, taken for a variety of real and imaginary complaints. Enrico Caruso always insisted that he be supplied with aspirin by his impresarios, and Franz Kafka explained to his fiancée, Felice Bauer, that it was only thanks to aspirin that he was able to face the adversities of existence. Aspirin for every little ache and pain: The good tolerability of the medicine withstood all the unreasonable uses to which it was put. The mechanism by which aspirin works was explained only late in the twentieth century, and furthermore, the effect of suppressing platelet aggregation and the preventative effects that it has been shown to have with respect to heart attack and stroke have given this wonder drug, even at the venerable age of one hundred, a new lease on life.[22]

## Alkaloids of Ergot

During the Middle Ages, many people died after eating contaminated rye bread. The contaminant was a fungus that grew on the ears of rye. Its name is *Claviceps purpurea*. At times the deaths from *Claviceps* reached epidemic proportions. Conspicuous was that the victims suffered from a gangrene of the limbs. At the end of the eleventh century this disease was such a problem that a religious order was created for the care of its victims. The patron saint of the order was St. Anthony, and consequently, the disease caused by *Claviceps purpurea* was often called "Saint Anthony's fire." The cause of the disease was not discovered until the seventeenth century. Since that time, the incidence of the disease has been restricted to isolated cases by the avoidance of contaminated grain.

Before the disease-causing effects of *Claviceps purpurea* or of *Secale cornutum* were known, there was already a drug derived from the fungus in use for accelerating contractions in childbirth. Such a use is mentioned by Adam Lonicer in his herbal published in 1582. But the dangers of such treatment were also known. Contractions that were too strong could lead to tearing of the uterine wall or to stillbirths. By the beginning of the nineteenth century, *Secale cornutum* was recommended only for control of postoperative bleeding.

In the years following, chemical and pharmacological research on alkaloids of ergot[23] was concentrated in England and Switzerland. George Berger and Francis Carr, of the Wellcome Research Laboratories, standardized ergot extract, which their firm sold for therapeutic and experimental purposes. In 1905, they reported the isolation of a complex of pure alkaloids, which they dubbed ergotoxin. Other researchers showed in 1943 that this complex was a mixture of ergocornine, ergocristine, and ergocryptine. Henry Dale, the Nestor of English pharmacology, developed bioassays by means of which the effects of the ergot alkaloids could be measured. Especially important was the suggestion that ergotoxin not only could promote contraction of smooth muscles (uterus, arteries, pupils), but could have the opposite effect on blood vessels when it was employed as an antagonist to adrenaline.[24]

From 1917 on, the Sandoz company, in Basel, became the driving force behind research into ergot alkaloids. In that year Arthur Stoll, a student of Richard Willstätter, joined the firm. Willstätter was one of the leading chemists of the beginning of the twentieth century.[25] It is thanks to him that chlorophyll was isolated and its structure explicated. For this he received the Nobel Prize in 1915. Arthur Stoll was one of his most brilliant students. The buildup of drug research at Sandoz began with research on ergot alkaloids. One year after taking up work in Basel, Stoll had already described ergotamine, the first pure ergot alkaloid, though it was another thirty-three years before its structure was understood. It was not until 1961 that Albert Hofmann and his colleagues, also at Sandoz, succeeded in a complete synthesis of this complex alkaloid. Out of research into ergot, whose medical applications seemed at first rather limited, came techniques of hypertension therapy, treatment of migraine, and, we think here of bromocryptine, of the treatment of tremors associated with Parkinson's disease and research into hallucinogenic mechanisms (lysergic acid derivatives).[26] The ergot alkaloids serve as an example of how consistent work on a naturally occurring substance through a dialogue between chemists and biologists can lead to the development of an entire area of drug research, in the present case with application to gynecology, hypertension, migraine therapy, and neurology.

These examples should content us for now. They show that there are comprehensible connections between modern drug research as we know it today and the groping beginnings of the knowledge of treating illness. They show at least that the production of extracts and mixtures of known effectiveness came at the beginning of industrial pharmacology and that the description of their components, their total synthesis, and subsequent synthetic modification in partnership with experimental pharmacology characterized the continuing progress of drug research in our century. In the following chapter we shall attempt to sketch the development of experimental pharmacology.

Figure 3.1. Filtering a sea onion. Page from an Arabic Dioskurides manuscript, 1222. Baghdad school. (Source: The Walters Art Gallery, Baltimore.)

# 3
/\/\/\/\/\

# Drug Research: Merging Experimental Pharmacology and Chemistry

## Emancipation Movements

Before the French Revolution there were emancipation movements already underway in many segments of French, and European, society. The professions, which until then had been reserved for the high bourgeoisie and nobility, were suddenly open to craftsmen and farmers, the sons of workers from lower social strata. For science, such an approach represented a new attitude toward experimentation. If science (philosophy, which represented erudition based primarily on traditional sources) and the crafts had been separated throughout the entire Middle Ages, now they entered a period of rapprochement. Therewith came the unprejudiced observation of nature, the acquisition of knowledge by means of experimentation, and the sober interpretation of phenomena free of pure speculation, which until then had been remarkably free from the constraints of observation or experiment. With the French Revolution there began in Europe a broad movement, which gradually encompassed all nations and societies of Europe, with the idea that the meaning of the world and the intellectual and spiritual appropriation of its manifestations should no longer be left to a few privileged social strata but should be a matter of concern to a wider circle. In Germany this attitude became prevalent in the period around 1830 to 1850. This general radical movement led to fundamental changes in medicine and pharmacology. At this time, anatomy, developmental biology, and physiology made significant strides. Whereas in the past, anatomical knowledge and teaching had been determined by the extrapolation of ancient or medieval sources, enlarged by current speculations, now, at the beginning of the nineteenth century, the scientist's own researches, done with his own hands, came to the fore. Whoever wanted to study anatomy had to dissect, observe, describe, classify, and in interpreting his findings had to remain in the framework of what could be clearly understood and verified. In physiology, animal experimentation took the prominent place it holds to this day, and the study of pure, or at least standardized, mixtures of plant products obtained in this period followed the pattern laid down by physiology. The advocates of this new experimental medicine came in large part from the social stratum occupied by craftsmen. François Magendie, the first experimental pharmacologist, and his contemporary François Broussais, whose orientation was more in the direction of anatomy, were sons of surgeons, a profession that unlike that of physician was at the time considered a craft. Magendie's most famous student, Claude Bernard (1813–1878), was the son of a viticulturist. Johannes Müller (1801–1858), the Nestor of German physiology, came from a family of laborers, and one of the first German pharmacologists, Carl Philipp Falck (1816–1880), was the son of a craftsman. Leopold Auenbrugger (1772–1809) was an innkeeper's son; he carried out his first percussion experiments on jugs filled with water to different levels and then went on to enrich diagnostic medicine with one of its most craftsmanly bases, thoracic percussion (tapping the chest).[1]

Medicine became a concrete subject and began to orient itself toward experience and observation and to pay less attention to abstract theories and speculation. It was only later in the twentieth century that medical science, above all molecular biology, remembered that before any experiment is undertaken there should be a speculative idea of the nature of reality behind it. Either this idea will be compatible with the experimental results, or else the experiment will prove it wrong, will "falsify" it. In its most creative form, science is frequently a dialogue between the intuitive and the analytic and experimental.

## The Republic of Cells

Other circumstances affected the science and practice of medicine. The political idea of the republic and its organs was carried over to biology. A state whose free and autonomous citizens are no longer subservient to it but who determine it through deciding the meaning of their own lives and the tasks they will undertake became the model of the human organism. Individual organs and cells were now the objects of scientific study and observation. This marked the beginning of what later came to be called "reduc-tionism" and which has been criticized especially in our era, particularly in connection with molecular medicine. The era of cellular biology, repre-sented in Germany by Johannes Müller and J.F. Sobernheim (1803–1846), had dawned. It reached a peak on the one hand in the cellular pathology of Rudolf Virchow and on the other in the molding of physiology and phar-macology into disciplines that more and more placed isolated organs, then isolated cells, and finally subcellular structures, at the center of their obser-vations. France was initially the leader on the path to this new medicine. François Magendie and Claude Bernard, teacher and student, were the founders of a pharmacology based on physiology. In Germany there fol-lowed, with a delay of about thirty years, Rudolf Buchheim and Oswald Schmiedeberg.

François Magendie was the first French physician to emancipate him-self from all scientific opinions that could not be supported by observation or experimental findings. He believed firmly that the laws of physics and chemistry held as well in the domain of living organisms as in the inani-mate world. His goal, as he put it, was to return physiology to "positive facts." He compared himself to a collector who noted facts without putting a value on them. In his study of pharmaceuticals he preferred to begin with drugs that were already in use in treating disease, but he always at-tempted to give an account of the active ingredients in the prevailing preparations before describing their pharmacological effects. The changes introduced into physiology by Magendie can be described as follows: He was the first to develop what could be described as a systematic technique for animal experimentation. He believed in a sober and precise descrip-tion of animal somatic functions. For him animal experimentation was a necessary method for testing the effectiveness of new substances. And fi-nally, he wanted to base the use of drugs on *people* on already proven ef-fects in animals. Moreover, his experiments on the effect of strychnine and emetine represent serious excursions into experimental pharmacology.

Claude Bernard continued the work of his teacher. In the way he posed questions he was closer in spirit to pharmacology. His particular interest was in poisons, but also in substances that on the basis of empirical evidence were already being used as medicines. He was involved to a particularly great extent with the effects and symptoms of nicotine poisoning. More strongly than his teacher, Claude Bernard was involved in a dialogue among observation, hypothesis, and experiment.[2] Like Magendie he emphasized the chemical and physiological basis of somatic functions, without going so far as to express them in mathematical formulas or place them in numerical categories. For that the time was not yet ripe. Claude Bernard was, however, already concerned with questions of absorption, distribution, kinetics, and metabolism of drugs, and he anticipated a number of questions that medicine was unable to answer until much later. For example, he posed the question of whether the controlled use of curare might make it easier to perform operations under anesthesia. In an analogous fashion he dealt with the question of using strychnine in suitable dosages for overcoming muscular paralysis. Neither curare nor strychnine fulfilled the therapeutic requirements that Bernard had in mind. As pharmacological and therapeutic methods, however, many of his ideas continue to be relevant. The development and introduction of muscle relaxants in anesthesiology in surgery illuminate this point, as does the use of cholinergic substances for treating a particular form of chronic muscular weakness: myasthenia gravis. Cholinergic compounds resemble the natural neurotransmitter acetylcholine, which affects muscular contraction.

## Effect and Function:
## The Dictatorship of the Experiment

The work of Magendie and Bernard reverberated in France and later in the German-speaking world as well. Given the large number of scientists involved, it is difficult to single out individual contributions to the evolution of experimental pharmacology. If Rudolf Buchheim is named here as the founder of this discipline in Germany, it is on account of the fact that he was the first to build a laboratory for experimental pharmacology.[3] This memorable deed took place in the German-language University of Dorpat, in Estonia, to which the physician Buchheim had been called as a young man to a professorship of pharmacology, dietetics, and the history of medicine. Buchheim emancipated pharmacology from physiology by giving drugs and their characterization his particular attention. The methodology and vocabulary of the new discipline remained those of physiology, but it was no longer somatic function as such that was the goal of experimental investigation, but the changes such functions underwent under the influence of toxins and drugs. Buchheim was also the first to decouple the study of pharmacology from therapeutic practice. What he described was the effect of substances on healthy experimental animals or on isolated organs or cells, not on patients. Since experimental conditions in animal research can be stricter and more invasive than in research on human subjects, the effects of drugs could be outlined in greater detail and accuracy, and thus drugs could be described independently of patients and disease. In this lay considerable advantages: the possibility of scientific precision, the quantification of dosage and effects, and the characterization of toxicity. But in this there also lay the danger that pharmacology would distance itself from therapeutic issues. Indeed, it succumbed to this danger, and only in the twentieth century did pharmacology return to its origins, the treatment of disease. If one wishes to understand the resistance that the young discipline of pharmacology was facing at that time, one needs to understand the general situation of medicine in Germany in the middle of the nineteenth century. In 1842, C.A. Wunderlich and W. Roser published the first volume of their new journal *Physiologische Heilkunde* (Physiological Medicine). They programmatically rejected all traditional cures that bore an a priori, unproved, classification. They likewise set themselves against the ideas of illness going back to Thomas Sydenham and against the ontological notion of disease, viewing illness as a functional deviation from the norm, which in every case bore individual features and therefore required individual treatment.[4] They likewise found unacceptable an empirical selection of drugs organized according to artificially created notions of illness, so-called nosological entities. According to the ideas of physiological medicine put forth by Wunderlich and his followers, a medicine should be put into use only after it has been established on the basis of animal experiments that it can in the therapeutically desired way influence and correct the function that has been adversely affected by illness. And indeed, under the influence of this new and rigorous conception eighty percent of the medicines used up to the middle of the nineteenth century disappeared from the pharmacopoeias. Physicians suddenly found themselves with very few medicines, and in the new, more stringently scientifically oriented, medicine there emanated a therapeutic nihilism, of which the Viennese school was a prime exemplar.

## Oswald Schmiedeberg: The Creation of Pharmacology

It was the historic achievement of Oswald Schmiedeberg (1838–1921) to continue on the path to a scientifically based therapy and to a clinical use of drugs that could always be compared with experimental findings. In so doing, Schmiedeberg created the conceptual and experimental framework of modern pharmacology (Figure 3.2). Schmiedeberg studied in Dorpat and completed his doctorate under the direction of Rudolf Buchheim.

**Figure 3.2.** Oswald Schmiedeberg (1838–1921) is considered the founder of modern pharmacology. In 1871 he was appointed to the newly established chair in pharmacology at the University of Strasbourg, where he remained until 1918. Under Schmiedeberg's directorship the Strasbourg institute became a leading center for research and education.

When Buchheim was called to a professorship in Giessen, the young Schmiedeberg took over the reins of the institute in Dorpat founded by Buchheim. In 1872 he went to Strasbourg to fill the chair of pharmacology at the newly created Imperial University. From then until 1918, the year in which Alsace returned to France, Schmiedeberg not only built a model institute in Strasbourg, but he created, together with his students, some of whom were connected with this institute over most of the next forty-six years, the subject of pharmacology as we know it today. One of the main focuses of his work was the difficult chemistry of digitalis glycosides. Schmiedeberg and his colleagues were the first to isolate pure digitoxin. But the hallmark of the Strasbourg institute was not only chemistry, but much more the systematic description of the effects of drugs on animals or on isolated organs. Dosage–effect relationships and other obvious categories of modern drug research come directly or indirectly from Schmiedeberg's research group.[5]

The close connection between biochemistry and pharmacology found its expression in research into drug metabolism. Among the interests of the Strasbourg institute was the detoxification of drugs by glucuronic acid, but there was also important work done on the synthesis of the components of cartilage, hyaluronic acid, and the relation of this acid to chondroitin sulfate, collagen, and amyloid. In the age of molecular biology it should not be left unmentioned that it was Schmiedeberg's institute that was the first to obtain a pure nucleic acid.

Schmiedeberg also created modern pharmacology in the sense that his students gradually assumed all the newly created chairs of pharmacology in the German-speaking world and beyond. From his school came also the first prominent industrial pharmacologists, such as Heinrich Dreser (Bayer) and Edwin Stanton Faust (Ciba). Two traditions important to drug research were united in Oswald Schmiedeberg's Strasbourg institute: First, going back to Sertürner, Pelletier, Caventou, and others was the isolation of pure active substances from complex mixtures, and second, an ever more precise and reductionistic description of physiological function, which began with Magendie and Claude Bernard and was refined by Buchheim, Schmiedeberg, and his school. Two further prerequisites were thereby created for the formation of industrial pharmacology. A further prerequisite lay in the development of synthetic chemistry and the concept of chemotherapy, which historically developed from dye chemistry. This will be the subject of the next section.[6]

## The Origins of Heterocyclic Chemistry

By about 1860 the theoretical groundwork had been laid for the creation of a chemical industry—as opposed to a chemical craft or trade. Avogadro's atomic hypothesis was confirmed and a periodic table of the elements established. That is, chemistry had developed a theory that allowed it to organize the elements according to atomic weight and valence, and thereby to predict the existence of new, not yet discovered, elements. There was also a theory of acids and bases. In 1865 August Kekulé formulated his pioneering theory of the structure of aromatic organic molecules.[7] The name "aromatic" comes from the fact that a number of these compounds, which were first found in plants, were in fact aromatic.[8] It was later discovered that such compounds contain less hydrogen than would be predicted by the molecular formula for aliphatic hydrocarbons, namely $C_nH_{2n+2}$. Turpentine, for example, an oil distilled from pine, has the molecular formula $C_{10}H_{16}$. For xylene, a distillate of the pine *Pinus maritimus*, the composition is $C_8H_{10}$. Eventually, a number of oils were obtained from the redistillation of coal tar, from which compounds with similar properties were crystallized. Naphthalene was one of the first and most important compounds of this sort, and over time a large number of "aromatic" substances derived from coal and tar became known. No less a figure than Justus von Liebig expressed doubt as to the value of too intense an interest in such substances derived by "destructive distillation." In his opinion, these compounds had nothing to do with "organized," by which he meant *living*, nature. The world of plant and animal chemistry should remain the central concern of chemistry, and the most important questions in chemistry had little to do with the solution to a problem of numerical relationships. Nonetheless, the number of such compounds with missing hydrogen atoms obtained through "destructive distillation" continued to grow, for coal tar was the first great waste product of the dawning industrial age, and it proved to be a gold mine for chemists searching for new compounds. In 1862, as he was writing a textbook, August Kekulé experienced a vision that intuitively solved the problem of the missing hydrogen atoms.

## The Vision of August Kekulé

Many of the contradictions present before Kekulé in the empirical molecular formulas and structural theory of hydrocarbons could be solved by hypothesizing the existence of ring-shaped molecules. However, it took von Kekulé three years before this intuitive inspiration was shaped into the benzene ring theory, to which important contributions were made by others as well, for example, Wilhelm Körner. It was Körner who proposed the terminology "ortho," "meta," and "para" for the possible placements of pairs of substituents in the benzene ring and who with these different placements solved, together with Kekulé and his colleagues, the problem of isomeric bonds in the ring system. Kekulé's effectiveness not only as an intuitive and theoretically brilliant scientist, but also as a teacher, is shown by the fact that three of the first five Nobel Prizes for chemistry went to his students: the Dutchman Jakobus van't Hoff (1901), Emil Fischer (1902), and Adolf von Baeyer (1905). To the first of these we owe the explication of cis–trans isomerism, while the second made fundamental contributions to the structures of sugars and proteins. Adolf von Baeyer explained the structure of indigo dye and was the first to synthesize this compound. Kekulé's benzene theory and the work of van't Hoff, who explained stereoisomerism and with the tetrahedron model gave a robust description of cis–trans isometry, set organic chemistry on a firm footing.

## A Kaleidoscope of Colors

The benzene theory gave a decisive impulse to research on coal tar derivatives, particularly dyes. William Henry Perkin, a student of August Wilhelm Hofmann, was professor of chemistry at the Royal College of Chemistry, in London, during the years 1845–1864. Hofmann had been working all his life on aniline. When a new aniline derivative with the molecular formula $C_{10}H_{13}N$ was discovered, Perkin attempted to use this compound in an oxidative reaction to produce quinine, with the molecular formula $C_{20}H_{24}O_2N_2$. With potassium chromate as oxidizing agent he obtained a product that defied characterization. But when he brought aniline itself into the reaction, a beautiful purple dye came out, which he named mauveine and which soon became a commercial success. Soon aniline became the starting point for a great number of new dyes. Organic chemistry was gripped in a veritable riot of colors. Aniline red, or fuchsine or magenta, was an important representative among the new dyes. A typical yield consisted of 250 grams of fuchsine from 850 grams of aniline, which in turn was derived from 1.4 kilograms of nitrobenzene. For the production of nitrobenzene, 1 kilogram of benzene was required, which in turn required the distillate from a ton of coal.[9] Further examples are aniline blue and aniline violet, the diazo compounds produced in 1858 in Hofmann's laboratory by Peter Griess, malachite green, and the yellow dye naphthazarine, which Carl Gräbe and Carl Liebermann accidentally produced during their search for a synthetic version of alizarine, the natural dye from madder root. Eventually, Adolf von Baeyer synthesized indigo, the "king" of dyes. The speed with which new dyes and synthetic processes that had just been discovered in the laboratory were patented and put into commercial production is reminiscent of today's dynamic in which newly identified genes or genetic products are taken over by industry. With respect to synthetic dyes, it was first France, then Germany, England, and Italy, and then again Germany that were at the center of this development, while today's analogue, the commercialization of biotechnology, is to be found primarily (ninety percent) in the United States.

## Dyes as Drugs: The Birth of Chemotherapy

The birth and development of dye chemistry had consequences for medicine as well, and in quite a particular way for the treatment of disease. The relationship between two apparently quite unrelated areas is one of the most remarkable examples of the continually repeated experience that "progress" in the broadest sense cannot be reckoned in advance and that the future cannot be predicted. For what historically unbiased reader would have expected that the isolation of dyes from coal tar, that the reappraisal of a previously despised waste product from the beginnings of the industrial era, could lead to a new class of medicines and to a revolutionary concept of the treatment of disease? And yet that is just how it was. It must be said, however, that the connection between dyes and chemotherapy, for that is the therapeutic concept about which we are talking, did not come about simply and directly. Two additional circumstances had to enter the picture in order that dye chemistry and medicine could bring forth chemotherapy.

### Diseases from Without

First, we must understand the nature of infectious diseases. The vague idea that diseases like plague, smallpox, or even malaria could be triggered by "outside agents," by external disease-causing influences, had been around a long time. In ancient times, for example in the writings of Hippocrates, there was already talk of "atmospheric" influences in connection with epidemics. In the Middle Ages the talk was of disease-causing miasmas (poisonous vapors), and the protective clothing of physicians and others who dealt with those sick with plague reveals that some kind of notion of infectious agents was suspected, agents that had attacked the sick and from which one could, with clothing and masks (to filter the air as it was breathed), protect oneself. "Malaria," the swamp fever, was, as its name reveals, believed to be caused by bad (*mal*) air (*aria*). What constituted this badness of the air? What was the material composition of these miasmas? What might be the nature of these atmospheric influences? To these questions no satisfactory answers were forthcoming. It was Giro-lamo Fracastoro (1478–1553) who first had the idea that certain diseases involving fever might be transmitted by invisible particles. In his time and given the means available to him, this idea could neither be proved nor re-futed. These particles, according to Fracastoro, arose spontaneously, but they could also reproduce and were capable of locomotion and could thereby traverse distances to transmit disease. Fracastoro's reflections come quite close intuitively to modern ideas of infection. His ideas show how greatly research is an active result of imagination and creative thought leading to the formation of hypotheses. Fracastoro's hypothesis was fully accepted neither in his lifetime nor over the next 250 years.[10] Scientific microbiology had to be created first through the work of Louis Pasteur before Fracastoro's hypothesis could be put to its first rigorous ex-amination (Figure 3.3). In the process, one part of the hypothesis was clearly refuted—"falsified," as Karl Popper would have put it.[11] Microor-ganisms do *not* arise spontaneously, but from their own kind, by cellular division in bacteria and, as was later discovered, by complex reproductive cycles in the case of parasites and viruses. However, the other basic tenets of Fracastoro's hypothesis still stand. The work of Louis Pasteur and his many students, above all the founding of medical bacteriology by Robert Koch and his school, confirmed Fracastoro's ideas in every detail (Figure 3.4). The first convincing practical application of the theory of infection, propounded and precisely proven in the nineteenth century by Koch and Pasteur, was that of disinfection during surgery, accomplished by the work of Joseph Lister (spraying of carbolic acid in the operating room, disinfec-tion of instruments). These methods resulted in a drastic reduction in peri- and postoperative infections. It marked the beginning of modern surgery and later was replaced by surgical asepsis—the systematic exclu-sion of microorganisms from operating rooms and instruments—and ex-panded by techniques of anesthesia. The founding of the modern doctrine of infection thus constituted the first prerequisite for the influence, intu-itively difficult to comprehend, of dye chemistry on medicine.

## Dye Affinities

The second prerequisite involved the naiveté, or perhaps ingenuity, of a young medical student who was looking at slices of tissue under a micro-scope in the laboratory of the famed anatomist Wilhelm Waldeyer, in Strasbourg. Many of the newly discovered dyes, but also older ones, were (and still are) used to color human and animal tissues so that when thin

Figure 3.3.   Louis Pasteur (1822–1895) is known principally for his germ theory of disease. (Source: CORBIS.)

Figure 3.4. Robert Koch (1843–1910) earned the gratitude of all mankind for his discovery of the causes of a number of infectious diseases, such as cholera, sleeping sickness, and tuberculosis. On the basis of his discoveries, these diseases can now be fought effectively. (Source: *Infectio*. Editiones Roche. F. Hoffmann–La Roche AG, Basel, p. 224. Universitätsbibliothek Basel, Portraitsammlung.)

sections are examined under the microscope, the cellular and subcellular structures are revealed more clearly. Among the hundreds, perhaps thousands, of young medical practitioners there was one who, looking at the riot of color under the microscope, wondered: Why do silver salts color nerve tissue? Why do certain tissue structures take up eosin or malachite green, while other structures are colored weakly or not at all? From such questions there gradually arose the idea that there are chemical affinities between particular dyes, on the one hand, and tissues—cells or cellular components—on the other. Paul Ehrlich generalized this question; that is, he posed it for chemical molecules in general and not just for dyes. But since there were no techniques available for studying colorless molecules, dyes were an ideal model for exploring the binding properties of biological structures and chemical compounds. At first, in histology, the subject of investigation was "dead," that is, "fixed," or denatured, proteins. The question of whether such selective binding and distribution of chemicals could also be demonstrated in living organisms led Paul Ehrlich to intra-vital staining, that is, the dyeing of living cells or tissues, or even the injection of dyes into laboratory animals. In this way he discovered that methylene blue colors nerve tissue, but that plasmodia—the malaria pathogens—also take up and bind this dye. Based on this "dye model," on the intravital staining with methylene blue and other dyes, Ehrlich developed over many years a theory of chemotherapy (Figure 3.5).

## Selective Binding: Effective Therapies

Ehrlich postulated that all cells, thus also microorganisms, carry on their surfaces, or in their interiors as well, certain receptors that are particular to themselves and that thus are not necessarily to be found in or on other types of cells. There exist corresponding dyes (or other substances) that bind with greater selectivity to nerve tissues, fat, muscle, epithelium,

Figure 3.5. Paul Ehrlich, born in Silesia, is considered the founder of chemotherapy. In 1907 he employed trypan red against trypanosomiasis. Two years later, he succeeded, in collaboration with his Japanese assistant Hata, in producing the arsenic preparation arsphenamine. Under the name Salvasan, it was the first effective drug against syphilis. Arsphenamine was displaced by penicillin only in the 1940s. (Source: *Infectio*. Editiones Roche. F. Hoffmann–La Roche AG, Basel, p. 61. Universitätsbibliothek Basel, Portraitsammlung.)

or even to certain parasites. In this connection Ehrlich spoke of organ-specific tropisms, and thus, for example, hemotropic, lipotropic, or even parasitotropic substances. The task of chemotherapy must then be to find such substances that are poisonous but that bind only to cells that one wishes to eliminate from the body of the patient. Such substances, administered from a therapeutic viewpoint, must have a "binding" component, and also a "poisoning" component that alters and kills off the bound structures, and these alone. Ehrlich wanted to begin with the available dyes and those becoming available in increasing numbers and systematically, step by step, bring their effects through chemical manipulation closer and closer to the requisite ideal. In this task animal experimentation, that is, the selective effect of dyes on disease pathogens in animals, should provide him the orientation for the chemical modifications. Here, then, the dialogue between chemistry and biology that became virtually proverbial for industrial drug research was for the first time expressly required. The importance of animal models for the development of chemotherapy can be seen in that Ehrlich suspended for ten years his research on the treatment of malaria with methylene blue because he had no plasmodium model available. Together with Paul Guttmann he had treated two malaria patients at the Moabit hospital, in Berlin, with methylene blue. Under this treatment the patients were cured. However, Ehrlich knew that he could not base a claim to a new type of treatment on two cases. Malaria was not a common disease in Berlin, so there did not seem to be favorable conditions for continuing clinical trials to prove the effectiveness of the dye. There were no animal models, and so Ehrlich decided to abandon this area for the time being and to return to it only under more favorable conditions. Ehrlich's path to chemotherapy may perhaps be more understandable if one considers that in the 1890s he was working on antibodies at Robert Koch's institute with Emil von Behring—or, as one then said, "antitoxins." The creation of a diphtheria antitoxin and the establishment of passive immunization, serum therapy, were achievements for which Paul Ehrlich and Emil von Behring were awarded the Nobel Prize for medicine.

Ehrlich himself was impressed by the selectivity of the antitoxins and called these molecules "magic bullets" because they acted exclusively parasitotropically and not organotropically—that is, on the parasites and not on the organs. "But we know," thus spoke Paul Ehrlich in 1909, "a host of infectious diseases, particularly those caused by protozoa, for which serum treatment is practicable either not al all or only over an extraordinarily long period of time. I mention here malaria in particular, the trypanosomal diseases, and perhaps a series of infections with spirilla. In these cases chemical means must come to our aid! That is, instead of serum therapy we should have chemotherapy!" The idea of a selectively acting chemotherapeutics was certainly characterized in its later formulation by experimental and clinical experience with serum therapy.

## Chemical "Magic Bullets"

To the extent that animal models were available, Ehrlich began with his colleagues to search systematically for substances that acted selectively against parasites. Hundreds of dyes were tried on a trypanosome model (mouse and rat) described in 1902 by von Laveran. Trypan red eventually proved to be a selective preparation against sleeping sickness in rodents. Unfortunately, it was ineffective against the analogous disease in humans. Ehrlich had indeed already indicated that dyes were merely ideal models for chemotherapy. When Laveran and Mesnil discovered a trypanocidal effect of arsenit in their trypanosome model, this discovery had at first only a theoretical significance. The sodium salt of arsenic acid was not considered a possible treatment because of its toxicity. Yet Breul and Thomas showed in 1905 that atoxyl (p-aminoarsenolbenzol) exhibited a genuine curative effect in trypanosome infections. At once Ehrlich began new researches with organic arsenic compounds, at first still with the trypanosome model; later, when it became possible to infect rabbits with the syphilis pathogen, with this first model of an experimental infection with *Treponema pallidum*. Arsphenamine, Ehrlich–Hata compound 606, was eventually selected as the best substance for human experiments. As "Salvarsan" this compound became the first effective treatment against syphilis and the first sweeping success of the new chemotherapy.

Paul Ehrlich had not just developed a new theory and created with this theory an experimental foundation with its own methodology; he had also put it to the test: the first effective chemotherapy, brought to market by the dye manufacturer Hoechst, with whom Ehrlich collaborated. We have to thank Paul Ehrlich for the theoretical foundations of chemotherapy: the idea of selective binding of dyes and drugs on chemoreceptors, the concept of binding (haptophorous) and toxic (toxophorous) groups on the drug molecule, the calculation of a therapeutic quotient (effective dosage divided by toxic dosage), and, as already discussed, the systematic dialogue between synthetic chemistry and biology. We have to thank him for Salvarsan as well, the first effective cure for syphilis.[12]

After Ehrlich, industrial research on chemotherapy went initially in two directions. One was the further development of organic arsenic compounds. The other was the buildup of the area of dyes. The first road led to further remedies against syphilis: tryparsamide and oxyphenarsin. On the dye front, eventually acriflavine and then suramin were developed—the first effective treatment against sleeping sickness. Over a longer period of time and by diverse paths the dye tradition led also to the sulfonamides.

## From Dyes to the First Sulfonamide

The dye chrysoidin—the hydrochloride of 2,4-diaminoazobenzol—which has been known since 1877, was first used for staining bacteria. In 1893, Eisenberg found that chrysoidin in a 1:10,000 solution inhibits the growth of gram-positive bacteria. In higher concentrations the dye also showed an inhibitory effect on gram-negative germs. However, the substance proved to be ineffective in vivo. For a time it acquired significance simply for the treatment of urinary tract infections. Mietzsch and Klarer produced the sulfonamide of chrysoidin in 1932, and in 1935 Gerhard Domagk found that this compound, prontosil, showed a hitherto unseen antibacterial activity in mice infected with streptococcus (Figure 3.6). However, in vitro, prontosil was inactive. Domagk was able to show that prontosil was capable not only of clinically curing mice infected with streptococcus or staphylococcus, but of eliminating the bacteria from the infected animals. Thus a bacteriological

**Chrysoidin**
(2,4-diaminoazobenzene)

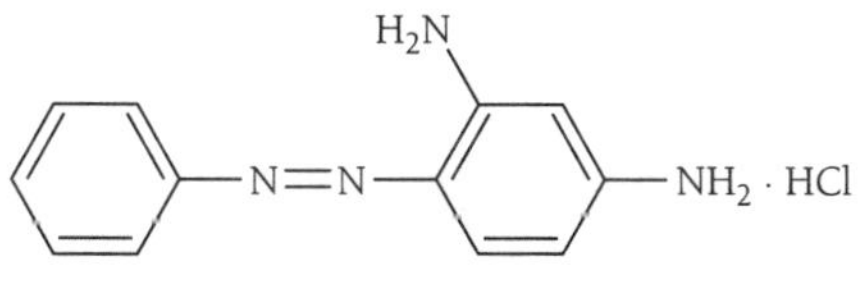

**Prontosil® "rubrum"**
(2,4-diaminoazobenzene-4'-sulfonamide, *sulfachrysoidin*)

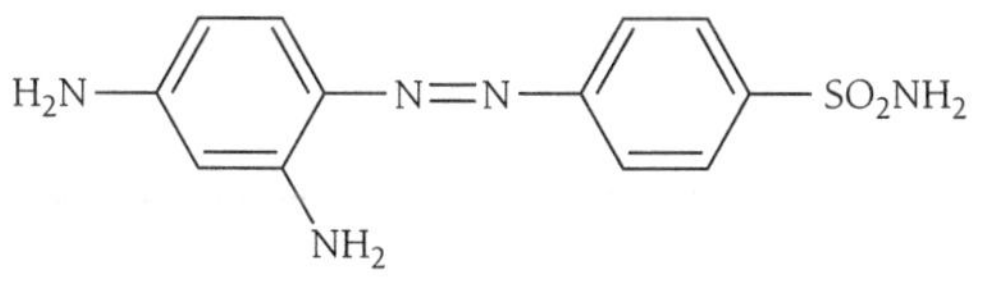

Figure 3.6.

cure could be achieved. In light of the ineffectiveness of the substance in vitro, naturally there arose the question of the mechanism of prontosil's action. The French husband and wife research team Trefouël already held the opinion in 1935 that sulfanilamide must be the active metabolite. Indeed, in the blood and urine of patients treated with prontosil a concentration of sulfanilamide was found corresponding exactly to the administered dosage of prontosil. From this, Colebrook and his colleagues concluded that sulfanilamide also inhibited the growth of bacteria in vitro. Domagk and his colleagues, on the other hand, were so firmly entrenched in their belief in chemotherapy based on dyes that they were unable at first to accept this theory. That the dye chrysoidin should be insignificant in its effect against bacteria seemed to contradict all experience. But that is how it was. In the end, the dye turned out to be merely the vehicle for the active compound, namely sulfanilamide. Its synthesis had been described already in 1908, unprotected by patent and available to anyone. Sulfanilamide, however, soon became the basis for the search for new, more effective and more tolerable, sulfanilamides. This search was, as we know today, extremely successful. Thus a dye became the instrument that made possible the discovery of a new class of drugs and led to the era of antibacterial chemotherapy.[13]

**Figure 4.1.** Model of a DNA double helix. Note the two strands of deoxyribose linked by phosphodiester bonds. They run in parallel forming shallow and deep "grooves." The complementary bases are stacked inside the molecule like the steps in a spiral staircase. (Courtesy of Genentech, Inc.)

# 4

## Technological Advances and a Paradigm Shift in Drug Research

We have seen how at the end of the nineteenth century various scientific traditions crossed paths: Analytic chemistry isolated pure and well-characterized active substances from traditional medicinal plants. Experimental pharmacology described the effects of these pure substances on animals in the categories of physiology. The isolation of dyes from coal tar led on the one hand to chemotherapy, and on the other it effected a decisive enlargement of the theoretical and practical bases of synthetic chemistry. Many cyclic and heterocyclic building blocks for the synthesis of new active substances are to be found in coal tar. Analytic chemistry, synthetic chemistry, pharmacology, and chemotherapy met in a narrow time period and together led to the founding of a new scientific discipline, modern drug research, in a likewise new institutional framework: the pharmaceutical industry.

In its now over one-hundred-year-old history, drug research has developed continually, benefiting from the progress that was made in the disciplines on which drug research was founded. Apart from this, however, drug research was continually confronted with "positive crises" that were solved with new technologies.

### Technological Advance: Positive Crises

After chemistry, chemotherapy, and pharmacology, which marked the beginning and first phase of industrial drug research, it was microbiology and the process of fermentation that gave drug research a new direction and bestowed on it a new, one might almost say dramatic, push forward. In 1929,

Alexander Fleming discovered penicillin, the metabolic product of a mold (*Penicillium notatum*) that had contaminated one of his staphylococcus cultures. In a large zone around the mold all of the staphylococcus cells had been dissolved, "lysed," as the bacteriologists say. The culture plate was completely clear and transparent. However, Fleming's ensuing description of the phenomenon attracted very little interest at the time. The observed phenomenon was simply another example in a long series of observations that had shown that many microorganisms produce substances which they excrete into their surroundings and that inhibit the growth of other microorganisms (Figure 4.2). This had begun in 1877 with a report by Pasteur and Joubert. These researchers had observed that cultures of *Bacillus anthracis* did not grow in the presence of "ordinary" bacteria. This phenomenon could also be reproduced in vivo. Mice into which anthrax bacilli had been injected died. But if they were injected with cultures of *Bacillus anthrax* contaminated with another microorganism, then they survived. In the following years, similar discoveries were made. An extract of *Pseudomonas pyocyanea*, which apparently contained an enzyme that was called pyocyanase and that could lyse other bacteria such as anthrax bacillus, corynebacterium, *Salmonella typhimurium*, and *Pasteurella pestis*, was

Figure 4.2.   In 1929 Alexander Fleming discovered that the growth of bacteria was inhibited in the presence of penicillium molds. The active substance involved was the natural antibiotic penicillin. (Alexander Fleming Laboratory Museum, St. Mary's NHS Trust, used with permission.)

used by Emmerich and Löw as a local antiseptic after they had shown systematically in animal experiments that there were no explicit side effects. The clinical application of the extract was recommended by such well-known bacteriologists as Escherich and extended to the treatment of conjunctivitis, abscesses, and angina caused by spirilla and accompanied by ulceration: Plaut–Vincent angina. The extract was even injected into the trachea or in cases of meningitis intrathecally, that is, into the spinal fluid. In 1912, in France, Rappin showed the inhibitory effect on the tubercle bacillus of cultures of B. subtilis, B. mesentericus, and B. megeterium. He also showed that the tuberculostatic effect was accomplished with filtered culture broth in which the strains were grown, and he named the active components "diastases." Up to 1929, a large number of similar observations were reported in the literature: In 1921 Lieske pointed out the bacteriolytic properties of actinomycetes. Gratia, an Italian researcher, described the growth inhibition of one strain of E. coli by another. Kimmelstiel and his colleagues found a strain of Bacillus mycoides that produces a lysin that later was used in the production of vaccines. This list is in no way complete. It should suffice, however, to make it understandable that the observation about which Alexander Fleming reported in 1929 and that was concerned with the antibacterial properties of a mold that had contaminated a staphylococcus culture should no longer cause a sensation: Too many observations of a similar type had already been made.

## Antibiosis: The Long Journey to Antibiotics

In 1889 the Frenchman Paul Vuillemin described the phenomenon of microbial antagonism as "antibiosis." The decision of Ernst Chain and Howard Florey to begin their search for antimicrobial agents obtained from molds with the penicillin discovered by Fleming turned out to be exceedingly fruitful. In a short time the Oxford group succeeded in isolating the active principle and in proving, first in animal experiments but soon also in human patients, its novel antibacterial effect. During the war, the American pharmaceutical industry was given the task of producing penicillin on a large scale. With the development of submerged cultures (as opposed to the at first more usual surface cultures), large quantities of penicillin could be produced to meet the demand for this medicine created largely by the great numbers of wounded. Penicillin, which soon after the war was produced in Europe (Grünenthal in Germany, Biochemie Kundl in Austria), began an unprecedented triumphal progress that has not yet ended. If the success of a medicine is measured in the number of years of life it has saved, then penicillin surely wins first place among all the medicines of the twentieth century. But penicillin was successful not only as a medicine: It was also the prototype of all modern antibiotics. The principle of antibiosis noted by Vuillemin had proved itself a brilliantly successful recipe. Now one could look for further antimicrobial agents. In 1944, the great American microbiologist Salomon Waksman isolated streptomycin from *Streptomyces griseus*. In 1939, Dubos had already isolated tyrothricin, an antibacterial agent with the character of a peptide, from *Bacillus brevis*. Waksman characterized these antimicrobial agents, themselves products of microorganisms, as "antibiotics."[1] In rapid succession and into the 1960s, almost all the important basic structures of antibiotics were discovered. Many of these new substances were also modified semisynthetically. Antibacterial therapy, and later antifungal therapy as well, had entered a new phase. Within two decades the pharmaceutical industry had developed an impressive armamentarium. Bacterial infections, which throughout human history had brought on dreaded epidemics and had taken more lives than any other disease, even more than all wars and natural catastrophes together, appeared to have been conquered. The victory, however, was not a permanent one. The genetic material of bacteria, that is, DNA, is, like the genetic material of all other cells, subject to continual change. Such mutations occur by chance. Usually, they have no structural or functional consequences for the affected bacterium. Often they result in a functional deficit, and sometimes they lead to a functional benefit. For example, bacteria can become resistant to antibiotics or chemotherapies. If such mutations occur in the absence of antibiotics, then they have no consequences. The resistant bacteria have no advantage over the population at large. However, should such a resistant mutation occur in the presence of an antibiotic, then the consequences are dramatic. Only the resistant bacterium survives, and it grows in the presence of the antibiotic into an entire colony of resistant bacteria. Such has occurred in places where antibiotics are omnipresent: in hospitals, medical practices, and farms (where feed containing antibiotics is used). The more carelessly one deals with antibiotics, the greater the danger of resistant strains of bacteria developing. Yet this problem only gradually became clear to medical practitioners and health agencies. A further complication was added: The development of resistance in bacteria does not follow only the ordinary pattern of mutation and selection. Bacteria can exchange genetic material among themselves—even across species barriers. Acquired resistance properties are thus not only passed on "vertically," from one resistant cell to a daughter cell, but they can also spread in the bacterial population "horizontally." This spreading out occurs through direct contact among bacteria or through the agency of bacterial viruses, so-called bacteriophages. Thus every antimicrobial therapy creates selective conditions under which resistant mutants can establish a foothold and multiply. By the 1970s, bacterial resistance had become a scientific and epidemiological subject of concern. However, until recently it seemed that physicians would be able to keep one step ahead, that the drug researchers and producers would be able to bring new antimicrobial agents to market faster than the pathogens could develop resistance. In recent years this confidence has been called somewhat into question. In hospitals multiresistant bacterial strains are now found rather frequently, bacteria, that is, that are resistant to a great variety of antibiotics. For a long time it was primarily the gram-negative bacteria such as the enterobacteria (intestinal bacteria) or the widespread pseudomonads that developed resistance. But over the last several years there have appeared on the scene gram-positive microorganisms, among which are to be found many dangerous pathogens like streptococci and pneumococci (agents of pneumonia and meningitis) with resistance to penicillin, and this has led to the breakdown of antibiotic therapies. The appearance of staphylococci resistant to all penicillins and in some cases even to the "ultima ratio," vancomycin, shows that medicine is perhaps approaching a situation in which we stand practically defenseless against the attack of bacterial pathogens.[2]

In the 1940s, the discovery and development of new antibiotics became an important goal of the pharmaceutical industry, a goal that has not been lost sight of. On the contrary, new organisms are constantly included in the search for antibiotics. Eventually, the spectrum of antimicrobial effects that were being sought was also significantly enlarged. While at first the search was for antibacterial agents, in the 1950s the emphasis was on antifungal agents.

Moreover, the question was ever more frequently heard whether microorganisms might be capable of producing substances with a completely different suite of effects. Today we know that this is indeed the case. The antihelmintic substance "ivermectin," a superior remedy against the tropical filariose; cyclosporine and FK 506, two new immunosuppressants; enzyme inhibitors like lovastatin; and many other active agents from microbes have shown over this period not only that microorganisms are a source of antibiotic agents, but that among their products are many compounds with interesting *pharmacological* properties. Microbiology, fermentation techniques, and isolation methods as well became in the course of several years part of the standard equipment of medical research. Thanks to these methods many new active substances have been discovered and produced. When molecular biology made possible in the mid 1940s the production of recombinant proteins, the methods of fermentation and cell culture received a new boost. What had itself begun as a revolution in pharmacological research in the 1940s now had at its disposal the technological tools that made possible the transition to an even more fundamental second revolution.[3]

## The Rise of Biochemistry: Enzymes and Receptors

Whereas the pharmacology at the time of Oswald Schmiedeberg and even the generation of his first successors was characterized primarily by physiological methods, from the middle of the 1940s it became ever more biochemically oriented. Louis Pasteur had demonstrated in the 1860s the existence of substances responsible for the process of fermentation. He called these, logically, ferments. In Pasteur's conception these ferments were still components of living cells. Later, Buchner, in Germany, was able to show that ferments, which were now called enzymes (Greek, *"within yeast"*), appear in the cell-free supernatant of yeast and function in the absence of living cells. A "living" process like fermentation could be reduced to chemical processes. In 1926 an enzyme, urease, was purified and crystallized for the first time. It was soon learned that enzymes are proteins, which in their catalytic effects often work with coenzymes, organic compounds of low molecular weight that facilitate particular chemical reactions such as oxidation, reduction, and hydroxylation. Coenzymes often have the character of vitamins: thus riboflavin (also known as vitamin B2), for example, which assists in hydrogen (proton) transport. It was soon discovered that vitamins are widespread biocatalysts that supply energy, catalyze the biosynthesis of macromolecules (proteins, fatty acids, fats), and can break down carbohydrates and other molecules with the release of energy; in short, vitamins serve as important participants in all biological processes. From the middle of the twentieth century enzymology developed to such an extent that intervention in the activity of these biocatalysts could be put to therapeutic use: Inhibitors of nucleotide synthesis were employed as cytostatic agents to inhibit cell reproduction. Substances already known, like the sulfonamides, were understood to be inhibitors of a biosynthetic capability indispensable to bacteria, namely, the synthesis of dihydrofolate. Chance observations led in new directions. In 1933, N.U. Meldrum and F.J. Roughton described an enzyme in red blood cells that catalyzed the reaction $CO_2 + H_2O \rightarrow H_2CO_3 \rightarrow 2 \rightleftarrows H^+ + HCO_3^-$. They called it carboanhydrase.[4] Many years later it was learned that sulfanilamide, the active form of prontosil, causes a hyperacidity of the blood that can be understood as a consequence of the inhibition of this enzyme. Then Davenport and Wilhelmi showed that carboanhydrase is present in large quantities in the kidneys. In 1949, W.B. Schwartz united these observations into the hypothesis that sulfanilamides must limit the supply of hydrogen ions in the kidneys through the inhibition of carboanhydrase. He tested this hypothesis on patients with chronic cardiac insufficiency and edema. With sulfanilamide there was an increased excretion of sodium (natriuresis) and with it an increased excretion of water. Sulfanilamide led the way to better carboanhydrase inhibitors such as acetolamid and later also to more effective diuretics, such as hydrochlorothiazide and furosemide.[5] In more recent times inhibitors of the coenzyme A-hydroxyglu-tamylreductase (HGR–CoA), substances that inhibit synthesis of cholesterol, have become important in the treatment of the familiar hypercholesterolemia. Inhibitors of angiotensin converting enzyme have demonstrated very convincing effectiveness in the treatment of various forms of hypertension. Enzyme inhibitors are also important in chemotherapy, for example in the treatment of HIV infection: The new protease inhibitors, together with inhibitors of reverse transcriptase that have been used for years, have appreciably improved the treatment of this viral infection.[6] All of today's drugs that are known to be safe and effective are directed at just under five hundred target molecules. More than two hundred of these targets are enzymes. The inhibition of enzyme activity, whether followed consciously or encountered serendipitously, represents one of the most successful strategies to date for finding new medicines. As of 1996, among the hundred most-used medicines in the world were thirty-six enzyme inhibitors, twenty-two blockers or activators of G-protein coupled receptors, twelve that interfere with ion channels, nine that bind to nuclear hormone receptors. The remaining twenty-one are split among various other biochemical actions. New technologies have been decisive for the development of enzymology, above all chromatography, spectroscopy, and the use of radioactive isotopes (Figures 4.3 and 4.4).[7]

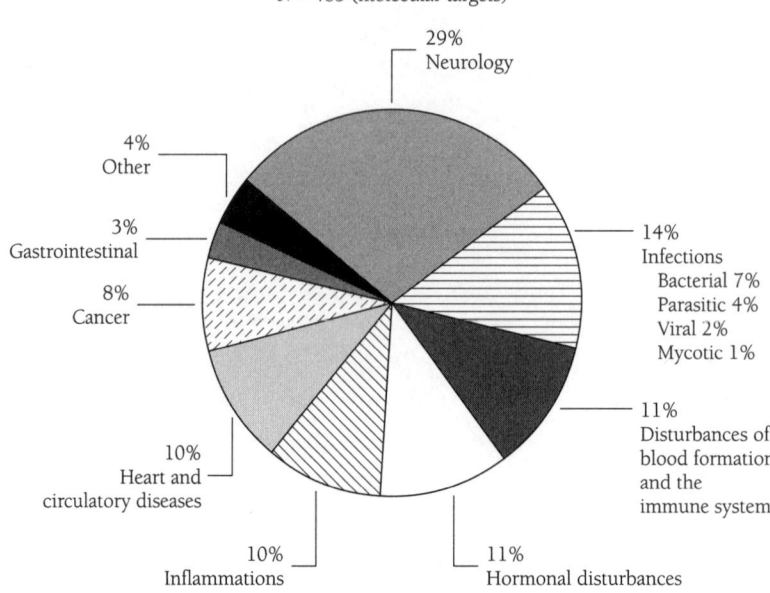

**Molecular Targets of Drug Therapy:
Classification According to Therapeutic Criteria**

N = 483 (molecular targets)

29%
Neurology

4%
Other

3%
Gastrointestinal

8%
Cancer

10%
Heart and
circulatory diseases

10%
Inflammations

11%
Hormonal disturbances

14%
Infections
Bacterial 7%
Parasitic 4%
Viral 2%
Mycotic 1%

11%
Disturbances of
blood formation
and the
immune system

**Figure 4.3.** Based on a modern standard work of pharmacology, the molecular targets of all known drugs that have been characterized as safe and effective have been collected and classified according to their principal indications. (Hardman, J., et al., *The Pharmacological Basis of Therapeutics,* 9E (1996), McGraw-Hill, reproduced with permission of The McGraw-Hill Companies.)

## Two Concepts of "Receptor"

After the biocatalysts, the enzymes, receptors were and are particularly rewarding targets for attack by drugs. The notion of a receptor has a chemotherapeutic (P. Ehrlich) and physiological–pharmacological origin. Ehrlich's concept of a receptor described simply a cellular structure that showed a particular affinity for a dye or other active agent and therefore formed a stable bond with such substances. For Paul Ehrlich such receptors were cellular structures suited to bind certain substances. He gave little thought to the cellular function of these chemoreceptors. His interest was in receptors as selective docking stations for toxic substances. *Corpora non agunt nisi fixata*—that is, substances can have an effect only when they are bound. The binding was effected by chemoreceptors. Of course, only such receptors were of chemotherapeutic interest that were characteristic

**Molecular Targets of Drug Therapy:
Classification According to Biochemical Criteria**

N = 483 (molecular targets)

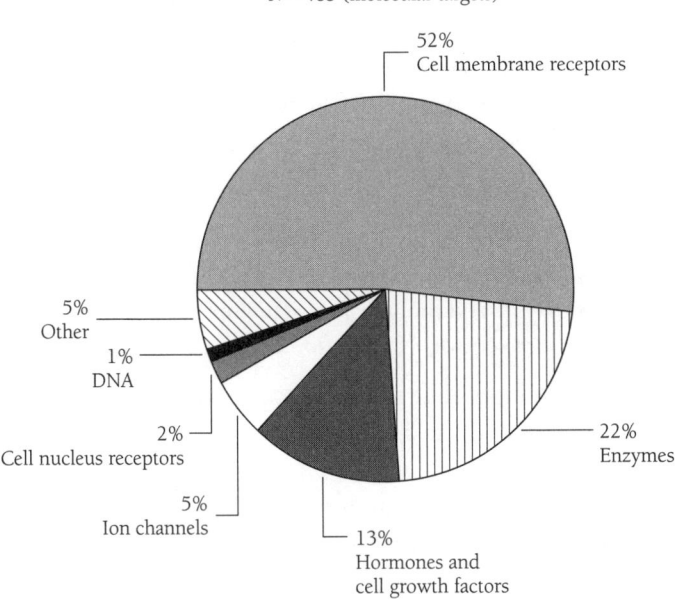

Figure 4.4. The drugs cited in Figure 4.3 arranged according to biochemical criteria. Receptors of the cell membrane and enzymes dominate all other categories as molecular targets for drugs. (Hardman, J., et al., *The Pharmacological Basis of Therapeutics*, 9E (1996), McGraw-Hill, reproduced with permission of The McGraw-Hill Companies.)

for a particular type of cell, preferably parasites or tumor cells that one wished to poison selectively and in this way rid the organism of them.

In contrast, the pharmacological concept of a receptor was the logical correlate of that of a neurotransmitter. As it turned out, to the extent that the neural stimulation of heart muscle or smooth, respectively striated, musculature was accomplished by particular chemically defined transmitters like acetylcholine, epinephrine, or norepinephrine, one posed the following question: By what structures is such stimulus mediated? In 1905, the English pharmacologist J.L. Langley introduced into pharmacology the concept, already mentioned by him in 1878, of specific receptors.[8] His model at the time was the neuronal end-plate in muscle and the excitatory, respectively paralyzing, effects of nicotine and curare on this structure. In 1948, R.P. Ahlquist published his now-famous article "A study of the adrenotropic receptors," in which he proposed the existence of two distinct types of adrenergic receptors. It is worth quoting here his formulation of two experimental results:

> "There are two distinct types of adrenotropic receptors as determined by their relative responsiveness to the series of racemic sympathicomimetic amines most closely structurally related to epinephrine. The *alpha* adrenotropic receptor is associated with most of the excitatory functions (vasoconstriction, and stimulation of the uterus, nictitating membrane, ureter and dilator pupillae) and one important inhibitory function (intestinal relaxation). The *beta* adrenotropic receptor is associated with most of the inhibitory functions (vasodilation, and inhibition of the uterine and bronchial musculature) and one excitatory function (myocardial stimulation)." (Ahlquist, 1948).[9]

From this classification there is a straight path to the various adrenergic blockers that were developed in the 1960s and that since then have achieved great importance in medical treatment. In the pharmacological definition, receptors were specific stimulus or signal receivers. Their function consisted not only in selective interaction with a specific signal, but also in its further transmission to the intracellular effector organelles. Seen historically, the concept of a receptor leads directly to the concept of intracellular signal transduction, which today has become one of the most intensively studied areas of cellular biology. Both research on receptors and the field of signal transduction were given additional impetus from molecular biology, because since the middle of the 1970s it has become possible to clone and express genes that encode these structures and thus to study the molecular structure of receptors directly. Similar possibilities exist for receptors of steroid hormones, thyroid hormones, and retinoids in the cell nucleus, as well as for ion channels and other proteins. The possibility of understanding enzymes, receptors, and ion channels functionally and to measure their activity, and then—mostly on an empirical basis—to search for substances that attack these biological targets with a certain selectivity, has enriched drug research enormously. One might say that the penetration of biochemical concepts into drug research is responsible for the drug revolution that in the 1950s and 1960s produced an abundance of new medicines, among which we single out psychopharmaceuticals, beta-blockers, calcium antagonists, diuretics, new anesthetics, and anti-inflammatory preparations.

Since the mid-1970s drug research, through molecular biology, has again been given a technological boost that has quickly gained in strength. It has already deeply altered this applied scientific discipline, and it will change even more in the coming years. At first, the changes were limited to the synthesis of proteins by genetic engineering. In this way, since human genes (more precisely, the transcripts of human mRNAs which are called cDNAs) could be cloned and expressed in bacteria. Proteins that were difficult or impossible to obtain previously suddenly became available. Human insulin, human growth hormone, and alpha-interferon were the first such proteins to be therapeutically administered. This enrichment of the therapeutic palette did not yet represent a therapeutic revolution. Nonetheless, the number of recombinant proteins grew rapidly, gradually including completely new substances such as colony-stimulating factors, G-CSF, erythropoietin, GM-CSF, and interleukin-3. Further interferons like gamma-interferon, a host of interleukins, and thrombolytic enzymes became available through recombinant techniques. Today more than forty recombinant proteins have been approved by the FDA. To be sure, this number includes clinically employed monoclonal antibodies.[10] Every year new proteins are added to the list. Up to the year 2000 the number of new recombinant proteins and monoclonal antibodies coming on the market annually will be in the range of twelve to twenty. That is a considerable portion of the new medicines introduced each year. According to recent estimates between ten and fifteen thousand human genes code for soluble proteins.[11] Of course, few of these proteins might be used for medicines, perhaps only a few percent. But even if the number of soluble proteins with therapeutic application were only one percent (a conservative estimate), then there would be altogether 100 to 150 new medicinal substances based on proteins to find and produce. If one counts all the proteins that have been found so far and clinically proven, not counting twice the same substance produced by different firms and also not counting the monoclonal antibodies, then one can expect a tenfold increase in the current stock of clinically useful recombinant proteins. And as we have said, this is a conservative estimate. If three to five percent of all soluble proteins were to lead to medicines, then the number would be significantly higher. Moreover, cell-surface receptors can be made soluble, or "solubilized," and with genetic engineering coupled to soluble proteins like antibodies: A fusion protein from the TNF-$\alpha$ receptor and a heavy chain of antibodies already under development is such an example. Thus there is still much to expect from "classical" biotechnology.

The true change in the direction of drug research comes, however, from another direction. The human genome contains about 100,000 genes, perhaps a somewhat greater number. All characteristics specific to the species and the individual are contained in this genome, and thus the susceptibility to certain diseases such as asthma, osteoporosis, hypertension, diabetes mellitus, cancer, arteriosclerosis, among many others. If from modern textbooks of internal medicine, neurology, and psychiatry and several smaller specialties like ophthalmology we were to count the diseases which represent medically and commercially viable targets for drug research, we would count about one hundred such diseases. In the causation of all these diseases there is a mix of genetic and environmental factors in varying proportions. In the 1920s, Sewall Wright developed a formula for estimating the number of genes associated with a complex phenotype in laboratory animals (mice and rats). For spontaneous hypertension in the rat the number of participating genes is four. For another breed of rats with diabetes mellitus type II, the number of genes involved in the phenotype is six. According to the estimates of leading geneticists, in the case of human disease there are probably no more than ten genes substantially involved in a complex syndrome such as hypertension. We can assume, then, that about one thousand genes in our genome are involved in the common multifactorial diseases. It is not at all certain whether the proteins specified by these genes are good targets for drugs. But all or almost all of these proteins are bound to be parts of frequently overlapping signal pathways in which many, for example twenty to one hundred, proteins take part. If we—again conservatively—reason further that each "disease gene" "corresponds" to three to five proteins that represent suitable target molecules, then our calculations yield between three and five thousand potential target molecules for new medicines. That is about ten times the number of biochemical targets influenced by the drugs available today. The true number of molecular targets for new drugs could be significantly higher, since all of our assumptions are on the "conservative side."[12]

One sees that a truly new dimension is opening up, without having taken into account that a precise understanding of the genome (or many genomes) and the genes and control elements will also help us to understand and possibly influence events such as aging, for example, as well as processes of differentiation and the development of organs. Genome research, combinatorial chemistry, modern techniques of screening, and bioinformatics with high throughput will alter the face of drug research in the coming years.

## Paradigm Shift: Chemistry and Molecular Biology

Modern drug research arose at a time when analytic and synthetic chemistry had attained a high degree of maturity and were moving on the one hand toward the industrialization of the purification and characterization of traditional substances and on the other in the direction of dye chemistry and the chemotherapy that was developing under its influence. We have explained that at this time experimental pharmacology was also in the process of becoming an independent medical discipline emancipated from physiology. The combination of these scientific–technical impulses let to the foundation of drug research, which found its appropriate institutional setting in the pharmaceutical industry. The culture of the pharmaceutical industry was uniquely determined by chemistry. It remained so even under the influence of the above-described technological advances due to microbiology and fermentation as well as those from biochemistry. Even molecular biology, which gave birth to an entire industry, has altered nothing in this regard in any decisive way. At its core, pharmaceutical research is still chemical. It rests on experience and also on the assumption, later theoretically and experimentally supported, that vital processes can be described in chemical categories and that diseases can be described as measurable deviations from "normal" chemical processes. Such deviations can also lead to changes in the chemical makeup of somatic components and organs. From this perspective drug therapy merely represents the attempt to normalize a dynamic equilibrium and bodily makeup through the provision of definite chemical substances. We have given examples in previous sections to describe how this happens. The inclusion of hormones, that is, regulatory substances, in treatment, like the use of cytokines, which likewise perform signaling functions, represents merely the attempt to effect a change in the course of reactions through influencing the control mechanisms. These various drug therapies influence chemical control mechanisms in order to restore a chemical balance that has been disturbed. The fundamental character of pharmaceutical research will not be changed through such methods. One could go so far as to understand drug research and the treatment of diseases with drugs as an expression of a chemical paradigm in medicine.

It is well known that the idea of a scientific paradigm stems from Thomas Kuhn.[13] By this term Kuhn, himself a physicist, denoted a closed and contradiction-free system of experiences, hypotheses, and examples that characterize a scientific discipline and its modus operandi at a particular time. Only when new facts not reconcilable with the old worldview (the old paradigm) become known can a new paradigm arise (usually in the course of a generation), which for its part creates a new framework for conceptual and experimental activity within the discipline. This, in brief, is the idea of Kuhn's thesis. Now, medicine is not a science like physics or chemistry; even more, it has been understood over the last 150 years to be an interdisciplinary attempt, characterized by scientific methods, to understand human disease and from this understanding to derive methods of diagnosis and treatment. With this modifying limitation, however, we can apply the term "paradigm" to medicine as well, and indeed with all the more entitlement, as the "paradigm" in question can be generalized. The notion that diseases are caused by external agents, that one can classify and describe them according to the nature of these causative agents, and that the elimination of disease microorganisms also leads to recovery from the corresponding disease, is correct. However, it is not generalizable to the same extent as the thesis that diseases are the consequence of disturbed chemical balance, that diseases can be described in terms of biochemical deviations from the norm, that an effective treatment must consist in the reestablishment of the chemical processes disturbed by the disease, and that this should occur in many, if not in all, cases through the administration of chemically defined substances. Something like this might be described as a chemical paradigm for medicine, and it is indisputable that modern drug therapy down to the present day can be accommodated within this paradigm.[14]

The origins of drug research, then, are in a "chemical worldview," and the cultural characteristics of chemistry at the time when the chemical, and soon thereafter pharmaceutical, industry arose have left their mark upon the procedures and culture of this industry. This is an extremely rigorous culture of precision and objectivity, but also of hierarchical dependency, discipline, and subservience. The atmosphere in the successful chemical institutes at the beginning of the twentieth century has continued within the industry. It is, even if weakened and overlaid with other influences, noticeable even today.

If one compares this strict, hierarchical culture given to formalism and schematization of functional processes with the culture of the biotechnology industry, one is struck by how much the conditions under which science is practiced have changed in the past fifty to one hundred years. In contrast to chemistry, molecular biology stems from a primarily democratic, liberal, indeed libertarian, social order, in which formal hierarchies play a much smaller role, while on the other hand, personal development and freedom are more important than in the society of a century ago. As a force based in industry, chemistry was a phenomenon of the end of the nineteenth century, while in contrast, molecular biology, at least in its industrially based and shaped characteristics, is a typical child of the departing twentieth century. The chemical industry developed its strongest expression from the countries of continental Europe, particularly Germany and Switzerland, while modern biology arose primarily in the Anglo-Saxon countries.

## Information: A New Paradigm in Medicine

Molecular genetics is a concept capable of introducing a new paradigm shift in medicine. In the new *informational paradigm*, which will prove itself to be the most comprehensive ever for understanding, diagnosing, and treating disease, the notion of genetic information plays a central role.

The human genome contains $3 \times 10^9$ base pairs in each set of chromosomes. Germ cells contain one set of chromosomes, while somatic cells hold twice that number. In this genome there are estimated to be about one hundred thousand genes. These genes comprise roughly five percent of the genome. To a small extent, the remaining ninety-five percent of the genome contains control elements (binding locations for transcription factors) and almost exclusively redundant DNA sequences with no apparent function. The genome contains all the directions for construction and function for a particular individual, including the genetically programmed plan of development. Particular genetic changes lead to functional loss and/or alteration. Diseases are such functional disturbances. They come about through the interplay of a variety of genetic changes. One might express the content of an informational paradigm for medicine in the following tenets:

1. All life processes are governed by a genetic program that is laid down in the organism's DNA.

2. Disease results from the incompatibility of a genetic program with a particular environment. Genetic ("informational") alterations that lead to or contribute to sickness can be caused by loss, reinforcement, or corruption of genetic information. In the development of epidemiologically significant diseases there are several altered genes involved.

3. Knowledge of the genome and the functions that arise from it will make it possible to describe and quantify diseases and susceptibilities to disease as informational errors or deficits.

4. The treatment of a multifactorial disease must ideally consist in replacing information that is lacking or correcting information that is erroneous. This can occur at the level of DNA (gene therapy) or at the level

of proteins specified by the genome. The latter case will be the domain of modern drug therapy.

An informational paradigm in medicine makes information the central point of consideration and thereby establishes itself as the principal all-encompassing attempt at interpretation of biological processes and the diagnostic and therapeutic measures derived from this understanding.

Of course, from this point of view gene therapy would be the most logical course of treatment, as it is directed at root causes. It is an open question whether it is technically realizable to the point where a general therapeutic principle can be developed. However, in every case the phenomenological consequences of genetic errors will be influenced more selectively and with greater versatility than is possible today. In addition, drug therapy built on genome research will develop in the framework of this informational medical paradigm. Nonetheless, even the most modern drug therapies will always remain a part of chemistry and of a chemical way of looking at biological processes.

It is possible that the new drug research—which is being developed and which will be very strongly marked by biology, the idea of information, and thereby a new paradigmatic understanding—will seek a new institutional framework. Today the classical pharmaceutical industry has already lost the monopoly on the discovery and development of drugs that it held for almost a century. Where will the discovery of new drugs eventually find its appropriate institutional basis? We shall go into this question in the last chapter of the book.

## The Influence of Molecular Biology on Medicine

The scientific medicine of the twentieth century was characterized by two assumptions: first, by the complementary notions of form and function; second, by chemistry, that is, by the question of the material composition of organisms, organs, and cells. Of course, these categories are not directly comparable, since the question of form and function can also be posed on the molecular level. If despite this conceptual inexactitude we speak, following Thomas Kuhn, of a morphological (and complementary to it a physiological) paradigm and of a chemical paradigm in medicine, then this, too, has a historical basis. Anatomy was already well developed at the beginning of the nineteenth century.[15] It was possible to describe exactly the position, size, and form of human (and animal) organs. Thereby the prerequisites were in place for two further developments: First, the question of function, and with it the development of a physiological perspective; and second, the systematic description of morphological changes that accompany disease—pathological anatomy grew in the nineteenth century under the influence of Rudolf Virchow to an independent scientific discipline in medicine, a discipline, moreover, thanks to which we have the first complete and consistent theory of disease. Basic phenomena of the behavior of tissues and cells were described, the different aspects of infection, for example expansive and infiltrating growth, the formation of metastases, differentiation and dedifferentiation, and much more. For the formation of the concept of disease that is valid even today this point of view was decisive. Yet it also led directly to important applications, for example to hematological and histological diagnostics. Even today therapeutic choices depend crucially on the question of whether a tissue sample shows signs of malignancy. Older and more modern imaging techniques—x-ray pictures, nuclear resonance, computer-aided tomography—build on the knowledge of normal and pathological anatomy, and what was at first a purely morphological point of view has been supplemented by highly differentiated functional diagnostics. For treatment as well, the morphological–physiological perspective has not been without influence. For surgery an exact description of morphological conditions was a compelling requirement, and in internal medicine the morphological–physiological viewpoint had predominantly diagnostic consequences.

Already at the end of the nineteenth century the hitherto exclusively morphological and physiological approach in medicine was overlaid by a new point of view: Chemistry had by then developed into a mature science, and it was almost inevitable that it would be applied to medical problems. On the one hand, bodily components, cells, organs, and bodily fluids became ever more the objects of chemical analysis, and on the other, the functions of the body and its organs were also described in the categories of chemistry. Physiological chemistry developed into a science that at first was able to describe the basic vital processes—metabolism, breathing, circulation—but then also aspects of these basic processes in all their complex ramifications. With this the necessary prerequisite was also given for describing the effects of substances that entered the body from the outside.[16] At the end of the nineteenth century pharmacology became a scientific discipline within medicine. On the one hand, chemistry was able to describe and explain somatic functions within chemical categories. By this it created the basis for today's laboratory diagnostics, the systematic recording of particular biochemical findings and their ascription to normal or dis-rupted organic function. In this way, medicine acquired knowledge of the "chemical signatures" of diseases. On the other hand, however, chemistry also provided the tools to modify these functions. Chemotherapy arose to-ward the end of the nineteenth century. At the same time, chemists learned to produce drugs according to chemical and pharmacological cri-teria that were at first roughly defined but were continually refined over the course of the following decades. We have discussed this development in the previous chapter.

The great successes of medicine in the twentieth century rest on the effectiveness of these two approaches to medicine: the description of form and function on the one hand, and on chemistry's ability to explain and influence somatic functions on the other. Not all fundamental innovations in medicine can be fit into this scheme. The evolution of the concept of in-fection and the understanding of infectious diseases cannot be described from the standpoint of morphology, physiology, and chemistry alone. The specific relationship between host and parasite must be included here. Nonetheless, the infection model in medicine does not go far enough to derive a general theory of disease. Something similar could be said of psy-chiatry. The description and interpretation of psychological phenomena require their own methodology. Psychological disease cannot be explained with chemistry, morphology, and physiology alone. Yet in turn the psy-chological methodology is not suitable to serve as a general approach for constructing a theory of disease. And at the close of the twentieth century the concepts and methods of chemistry were again of decisive importance in the *treatment* of both groups of illnesses.

## The Molecular Foundations of Heredity

The rise of a new paradigm in biology and medicine began in 1944. A research group then working in New York at Rockefeller University— O. Avery, C.M. MacLeod, and M. McCarty—identified deoxyribonucleic acid, which had been discovered earlier by Friedrich Miescher, as the "hered-ity molecule."[17] The chemical nature of genes had been decoded. One of the most consequential discoveries of this century followed in 1953: The explication of the structure of deoxyribonucleic acid (DNA) by James Wat-son and Francis Crick.[18] From the structure of the double helix questions arose with a certain inevitability, questions that were soon answered (Fig-ure 4.5).[19] It was recognized that the doubling of DNA necessary for cellu-

**Deoxyribonucleic Acid (DNA) as Universal Carrier
of Heritable Information**

DNA: A long molecular strand with the four building blocks A, T, G, C

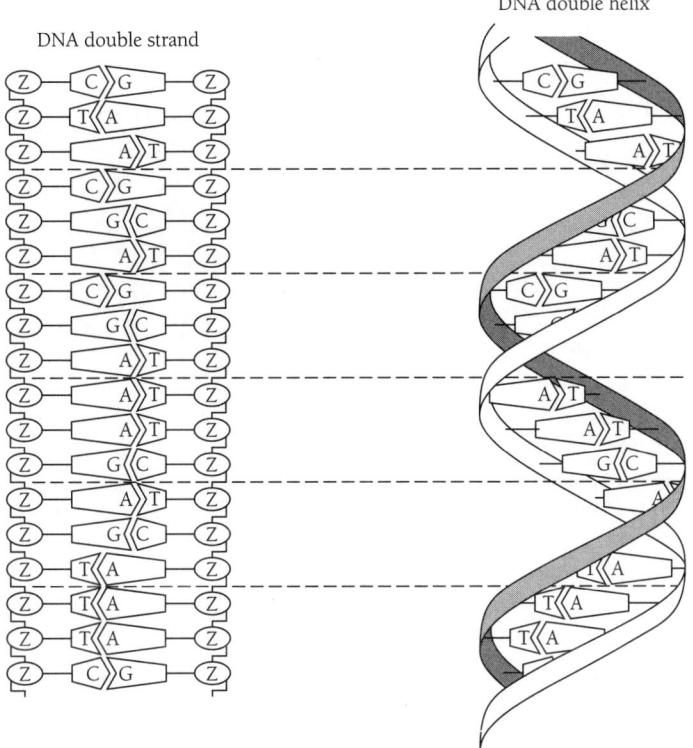

Human DNA: $3 \times 10^9$ building blocks
E. coli DNA: $3 \times 10^6$ building blocks

**Figure 4.5.**   Double-stranded DNA, shown first in linear format and then in the form of a double helix. Note the complementarity of the bases. The building blocks adenine and thymine on one side are always paired with guanine and cytosine on the other. The structure of a single strand determines completely the complementary structure of the other strand.

lar division was accomplished through copying the individual strands of DNA. This process was named "semiconservative replication." The enzyme responsible for this fundamental process, DNA polymerase, was isolated and characterized.[20] The equally important fundamental processes of transcription and translation were described, that is, the rewriting of DNA into ribonucleic acid (RNA) and the translation of RNA into proteins. Nature's language, the genetic code, was thus deciphered[21] (Figure 4.6). W. Arber,

**Genes Are Blueprints for Proteins**

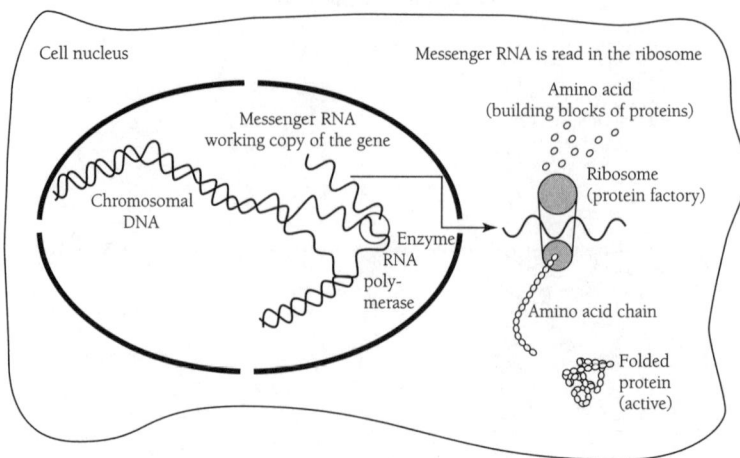

**Figure 4.6.** The genetic information contained in DNA is transcribed by RNA polymerase into RNA. Messenger RNA leaves the cellular nucleus and is transcribed into proteins in the ribosomes with the help of biochemical "ushers," the transfer RNAs (not shown). The ordering of the amino acids in the protein corresponds exactly to the arrangement of the "code words" (trinucleotides) in the messenger RNA.

D. Nathans, and H. Smith described restriction enzymes, with which DNA can be severed as if with scissors at particular loci,[22] and in 1967 various research groups found the ligases, which are enzymes that can recombine the cut-up genetic material. "Vectors" were identified, viral genomes or modifications of circular DNA and RNA molecules that can serve as "transport vehicles" for genetic material.[23] Finally, the difficult assignment of determining the sequences of nucleotides in DNA quickly found two solutions: one by F. Sanger[24] and another by A.M. Maxam and W. Gilbert.[25] With the development of a method for copying DNA rapidly and precisely, namely the polymerase chain reaction, or PCR (Ken Mullis and colleagues), the "toolbox" of genetic engineering was for the time being complete.[26] Of course, the history of molecular biology didn't end here. The complete sequence of various bacterial genomes and the genome of yeast were obtained. At least partial sequences of most cDNAs—the messenger RNA molecules in human cells that are rewritten by reverse transcriptase into DNA—are known, and the sequencing of the human genome is making rapid progress. About two hundred genes whose alteration triggers monogenetic diseases have been localized and described. These "disease genes" have been cloned, that is, isolated and reproduced. In many cases, as for example with the gene for cystic fibrosis or with the gene whose alteration causes muscular dystrophy, the functions were determined without which the disease ensues. In addition, genes have been found that are associated with diseases caused by many factors, diseases such as diabetes mellitus type II, obesity, hypertension, and a variety of forms of cancer. In this we are just at the beginning. One thing, however, is becoming continually clearer: Molecular biology, the science of the information that lies at the base of all life, is giving us a new, universally applicable access to medicine, a sort of master key. Diseases can be interpreted as information deficit, information excess, or as the effects of defective information. If health is the result of harmony or balance between a particular genome and the environment in which it finds itself, then diseases are disturbances that establish themselves when the information present in the genome does not permit an optimal accommodation to the surrounding environment.[27] Interestingly, these disturbances seem to occur in regular patterns.

## More Than the Sum of Its Parts: The Genome

Genes determine the structure of all proteins. However, the genome in its entirety does not contain only this structural information. It comprises as well all governing mechanisms that are responsible for the turning on and off of genes. Thus the genome determines the structure and form of an organism through the regulation of growth and development processes, to which the programmed death of cells, apoptosis, also belongs. Therefore, we must envisage genes and the genome not simply as information storehouses, but as self-regulating—in time and under variable environmental conditions—information systems. Lastly, we do not know how ten to twelve thousand genes organize themselves in such a way that, say, a fruit fly is created, and even less so how complicated organisms arise from a program that controls its own transformation. But we are beginning to understand how disturbances in the stock of information or in the retrieval and transmission of genetic information determine phenotypes, some of which have the character of diseases. Putting the notion of disease on the level of information and control, that is, from visible phenotype to causal genotype, has deep consequences for medicine. This holds as much for the understanding of diseases as for diagnosis and treatment. Here we collect once more the principal propositions of an informational–cybernetic paradigm in medicine:

1. The form and function of an organism derive from the information content of its DNA.

2. The genetic information of an organism also contains control elements for the transformation of the information into ordered spatial structures that change themselves over time according to a particular regime (self-regulating information system).[28]

3. Disease results from missing, overabundant, or defective information as well as from disruptions to inter- or intracellular transmission of information. Such information errors can be used in the diagnosis of diseases.

4. There are therapeutically useful relationships between diseases and the information content of a cell.

The technological possibilities coming out of molecular biology have already had a lasting effect on medicine. Until now, most innovations in diagnosis and in therapy have remained within the conceptual framework of morphology, physiology, and chemistry. However, it is ever more certain that far-reaching innovations and radical changes that will also alter the practice of medicine are beginning to emerge. We might express it as follows: Through knowledge of the human genome, medicine will be in a position to recognize and quantify diseases and susceptibilities early. Such estimates of individual risk will form a rational basis for a comprehensive program of prevention. Thus medicine will be in greater measure a practice of diagnosis and prevention. Furthermore, the knowledge of the structure, function, and interdependencies of genes will put us in a position to define potential points of attack for medicines in a more focused and selective way than has been possible hitherto. With this knowledge a new basis for drug therapy will be created. Gene therapy, too, will obtain a thematic context that today it does not yet have.[29]

## The Four-Phase Model

The influence of molecular biology on drug therapy, or more generally on therapeutic techniques, has grown gradually. When the methods of artificial recombination of DNA were mature enough that one could contemplate an industrial exploitation within the scope of drug therapy, people concentrated at first on the production of proteins whose therapeutic effects were already known, or at least foreseeable. Human growth hormone, human insulin, and thrombolytic enzymes like tissue plasminogen activator were the first proteins to fall into this category. We have denoted this very early phase as the *exploratory phase*, and today we may consider it to have ended. At about the same time as these initial steps were taken, the new methods were applied to obtain proteins either as yet unknown or known only by their biological activity. For this there were various techniques, which we shall not go into here in detail.[30] Only this: The purification itself of small quantities of a protein such as α-interferon permitted at that time the determination of the amino acid sequence or the decoding of partial sequences. Based on the amino acid sequence, the probable nucleotide sequence of messenger RNA could be determined. The genetic code is not quite precise: To each amino acid there is not only one, but three, possible base triplets. One speaks of the "degeneracy" of the code. Because this is so, the nucleotide sequences of messenger RNA can be determined only approximately from the sequence of amino acids in the corresponding protein. This situation puts a limit on the precision of genetic probes. Nevertheless, with such probes which correspond to the messenger RNA sequences, one can isolate the sought-after gene from cDNA gene banks and express it in suitable vectors. This happens with hybridization experiments. In this or a similar way the genes for a large number of cytokines were cloned and expressed in microorganisms or suitable mammalian cells such as Chinese hamster ovary cells (CHO). Thus molecular-biological techniques were decisive as much for the unambiguous identification of proteins that hitherto had been described only phenomenologically as for their production in large quantity and their therapeutic application (Figure 4.7). We have named this phase—to which we owe, among others, the interferons, many interleukins, erythropoietin, and colony-stimulating factors—the *biotechnology* phase. The palette of proteins that can be produced by genetic engineering technologies was significantly enriched by the fusion of genes that control the synthesis of fusion proteins in suitable vectors.

Georges Köhler and Cesar Milstein reported in 1975 on the hybridoma technique and the production of monoclonal antibodies.[31] At first these antibodies (Figure 4.8) were produced only in mice, but even after it had been learned how to fuse mouse lymphocytes with human myeloma cells and in this way to synthesize chimeric antibodies, the possibilities for therapeutic applications of monoclonal antibodies remained limited. The human immune system reacts to mouse antibodies with the formation of its own antimouse antibodies, which first of all lead to undesirable side effects via complement fixation and, secondly, can lead to a

## Cloning and Expression of the Human Alpha Interferon Gene in Coliform Bacteria

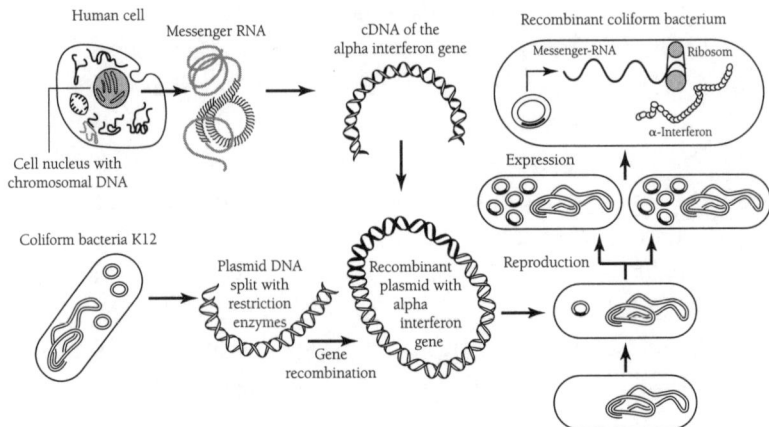

**Figure 4.7.** Messenger RNA is first transcribed into complementary DNA (cDNA). This cDNA is then inserted with the help of a restriction enzyme into a cleaved plasmid. The breaks are then sealed under the action of other enzymes, the ligases. The altered plasmid is then reinserted into a bacterium. The bacterium thus transformed now synthesizes certain proteins based on the code of the imported genes (interferon in this example).

neutralization of the therapeutic antibodies. This can proceed to the point where all biological effectiveness is lost. Only the possibility of replacing all the mouse sequences not belonging to antigen binding sites with human antibody sequences and leaving only the CDR (complementarity defining regions) would provide a far-reaching solution to the problem. "Humanized" antibodies were absorbed into the repertoire of genetic engineering technologies and are on the way to becoming important therapeutic tools (Figure 4.9).

All in all, genetic engineering has attained considerable therapeutic importance through the preparation of recombinant proteins, monoclonal antibodies, and fusion proteins, by which, for example, the receptor-binding part of a cytokine like TNF or interleukin-2 is fused with a heavy antibody chain. What we are dealing with here is far from being an experimental method; these are well-established techniques for the discovery and production of new types of medicines based on proteins. At the time of this writing more than seventy such medicines have been approved for the American market, many of which are also available in Europe (Table

**Monoclonal Antibodies from Hybridoma Cells**

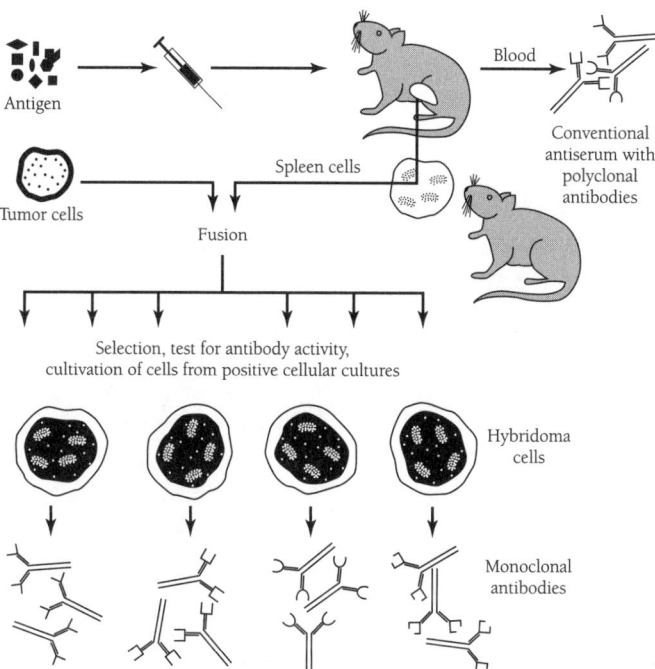

Antigen

Tumor cells

Spleen cells

Fusion

Blood

Conventional antiserum with polyclonal antibodies

Selection, test for antibody activity,
cultivation of cells from positive cellular cultures

Hybridoma cells

Monoclonal antibodies

**Figure 4.8.** Production of "monoclonal" antibodies. A mouse is immunized with a certain antigen. After several weeks, the B-lymphocytes produce antibodies against various chemical structures of the antigen. Each lymphocyte produces its own, very particular, antibody. A mix of these various antibodies is found in the immunized animal's blood. If the animal's spleen cells are fused with myeloma cells, then after thinning and cultivating individual fused cells, hybridomas—cell cultures that produce only a single type of antibody—are obtained.

4.1). About 250 new recombinant proteins and antibodies are presently in some phase of clinical testing. Various estimates indicate that around the beginning of the new millennium, in a few years, that is, about fourteen to twenty-four new recombinant proteins and antibodies will arrive on the market every year. Even if only the lower estimate is achieved, the number of recombinant medicines based on proteins would then about equal that of new low-molecular-weight substances also expected at this time. Many medicines obtained by genetic engineering, one might even say the greater part of this group, are novel substances. The usefulness of α-interferon against chronic hepatitis of types B and C and against lymphomas and

**Humanized Monoclonal Antibodies**

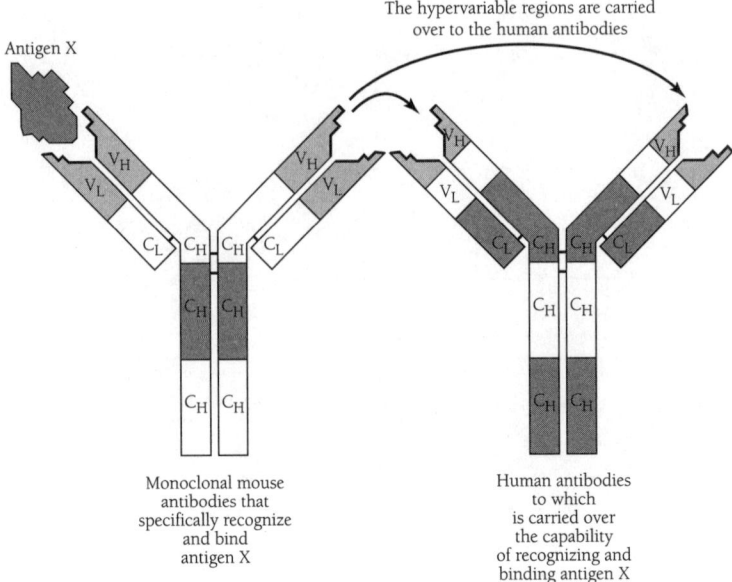

Figure 4.9.   Through gene-technological methods, the components of a mouse antibody responsible for binding the antigen are converted into a human antibody. In this way a "human" antibody is created that possesses the binding properties of the mouse antibody. This process is called the "humanization" of antibodies.

other malignancies is uncontested. Erythropoietin has attained an important position in the treatment of anemias caused by renal dysfunction. The colony-stimulating factors G-CSF and GM-CSF have taken a firm place in the chemotherapy of malignant tumors. They are administered alternately with regular chemotherapy in order to avoid the formerly feared loss of white blood cells (leukopenia). Tissue plasminogen activator (TPA) has in the meantime proven itself as a thrombolytic agent that is noticeably superior to old types of streptokinase and urokinase in the treatment of acute cardiac infarction. Moreover, this substance has also proved itself in the treatment of strokes caused by blood clots. When administered within the first three hours after such a stroke, thrombolytic therapy reduces the number of deaths and extent of long-term neurological damage. At present it is being tested whether such a positive effect can also be obtained if the medication is administered more than three hours after the event. A chimeric monoclonal antibody against fibrinogen receptor IIB/IIIA affords extensive protection against renewed blockage of coronary blood vessels

Table 4.1.  The most successful gene-technological medicines: Worldwide estimated sales.

| Recombinant protein (h = human) | Discovered/developed by | Country | Worldwide sales 1997 (estimated millions of U.S. dollars) |
|---|---|---|---|
| h-EPO | Amgen | USA | 3000 |
| h-Insulin | Genentech, Eli Lilly, Zymogenetics, NovoNordisk | USA/DK | 2400 |
| h-G-CSF | Amgen | USA | 1700 |
| h-Growth hormone | Genentech, Eli Lilly, NovoNordisk | USA/DK | 1600 |
| h-alpha interferon | University of Zurich/Biogen, Genentech/Roche, etc. | CH/USA | 1200 |
| Hepatitis B vaccine | Genentech, Biogen | USA | 1100 |
| h-β-Interferon | Chiron, Berlex, Biogen, Serono | USA, CH/Israel | 650 |
| h-TPA | Genentech | USA | 400 |

after their reopening and enlargement by angioplasty. In phase II clinical
studies, several monoclonal antibodies have shown good therapeutic effects against B-cell lymphomas (anti-CD28 from IDEC), while other antibodies against integrins and adhesion proteins appear to be effective
against inflammation and circulatory reactions induced by shock. A fusion
protein by which the soluble part of the TNFα receptor not anchored in
the membrane was fused with the heavy chain of an $IgG_2$ antibody neu-
tralizes TNFα and can possibly be used in the treatment of septic shock.
β-interferon has demonstrated at least partial effectiveness against multiple sclerosis, which up to now has been untreatable. This is only a short
list of examples that show that the initial phases of genetic engineering
have already begun to bear fruit.

Along with the direct application of proteins for therapy there have, of
course, also been attempts to manufacture proteins as pharmacological
targets through genetic engineering in order to use them experimentally as
models for the effects of small organic molecules. Many enzymes that
show promise of pharmacological interest and likewise receptors, such as
the various serotonin receptors, GABA receptors, and many others, have

been cloned. Now their molecular structures can be investigated. Above all, one can now with the help of combinatorial chemistry and other techniques of chemistry search for substances that interact in a particular way with these pharmacologically interesting proteins (receptors, enzymes, carrier proteins, proteins of the cytoskeleton, ion channels, etc.). We have named this phase of the influence of molecular biology on drug therapy the "pharmacological phase."[32] It has attained a special importance today through genome research. We shall return to it later.

Finally, there is a gene-technological fourth phase in the relationship between molecular biology and drug therapy. Defective or nonexistent genetic information can be replaced by techniques of gene therapy in such a way that the functional deficit can be alleviated at least temporarily. The strategies available for this will be discussed in the section on genetic engineering. Suffice it to say that the possibility of introducing genes into the germ line of animals and in this way producing transgenic mice, rats, and in increasing numbers other species represents a tremendous advance for the creation of disease models. Just as one can express particular genes in transgenic mice, it is also possible to inactivate individual genes or several genes in the germ line. Experimental animals in which such intervention has taken place are called "knockout" animals. Today transgenic mice as well as knockout mice play an ever more important role in drug research. In connection with genome research this technology is gaining even greater importance. If one could identify the genes whose absence or decrease in function contributes to so-called multifactorial diseases, than one could also "generate" experimental animals whose genotype and phenotype approach ever more closely the biological patterns of human disease. In this way the experimental models for investigation of medicines will merit a credibility and authenticity far superior to the animal models available today.

## Drug Therapy and Genome Research

Before we go into the influence of genome research on medicine in general and on drug research and therapy in particular, we wish to recall what genome research actually is. By "genome" we understand the entire DNA of an organism or a cell that is inherited by the daughter cells by semiconservative replication. In higher organisms this DNA is organized in chromosomes, which together essentially constitute the cell nucleus surrounded by a nuclear membrane. In somatic cells the genome exists in diploid form; that is, each gene and each DNA sequence exists in two versions, called alleles, of which one was inherited from the mother, the other from the father. During reduction division (meiosis) only the haploid chromosome set, that is, only one of two DNA double strands, is inherited. Mitochondrial DNA comes exclusively from the mother. It is not a component of the genome and will not be discussed here.

The progress that has been made in genetics, above all the technological innovations in molecular biology, has made it possible for many years now to determine the base sequences of DNA molecules and to copy arbitrary nucleic acid sequences with the polymerase chain reaction (PCR) method.

Thus "genome research" denotes the attempt to describe the structure and function of an organism based on its genetic information. This attempt comprises the following operational goals:

1. The construction of as many complete genetic maps as possible.
2. The sequencing of all the genes of an organism expressed at some point in its life.
3. An understanding of the function of these genes and of the mechanisms that lead to their expression at particular times and under particular spatial or biochemical conditions.
4. The knowledge of all uncoded parts of the genome and an understanding of their function (if any) for the three-dimensional structure of DNA (the chromosomes) and thereby for processes such as recombination, transcription, replication, and gene expression.

## Genetic Maps

The idea that these desiderata can be obtained most radically through decoding the entire nucleotide sequence of the human and other genomes led to the creation of the human genome project. In 1988 the then director of the National Institutes of Health, James Wyngaarden, called upon scientists and administrators to plan the Human Genome Project. In 1990 this program was officially inaugurated. By 1992 the first complete human genetic map was published, based on the order of microsatellite DNA in the human genome. Microsatellites consist of paired repetitions of two or more nucleotides, for example $(CA)_n$. The number of these repetitions varies from individual to individual. Microsatellites offer the advantage that they can be amplified by PCR. The genetic map published in 1992 by

Jean Weissenbach and his colleagues contained 5,262 such "markers" in 2,335 positions. The existence of such "place markers" separated by distances of less than 1 cM (centimorgan: 1 cM corresponds in man to $10^6$ base pairs) simplifies the location of genes responsible for particular phenotypes. For example, if a particular marker appears more frequently in the DNA of a patient with a genetic disease than in the general population, then this is an indication that the gene responsible for the disease might be located in the neighborhood of this marker. With such information one can go to a gene library and look up which genes lie in the vicinity of the marker that segregates with a given phenotype. In such situations one often is still left with a large number of genes, and one then employs various techniques to look for the "best candidates" among them.[33]

A genetic map of the mouse was constructed by E. Lander, W.F. Dietrich, and their colleagues. It contains 6,336 such markers distributed over the entire mouse genome. Such "physical" maps give a spatial scaffolding into which one can insert genetic data.[34]

The genetic maps described here are based on the analysis of gene libraries that are placed in YACs (yeast artificial chromosomes), cosmids, or other organisms (cells). Thus a cosmid will contain up to 45,000 base pairs of foreign DNA; in a YAC one can fit one to two million base pairs, thus one to 2 Mb. If the DNA of a genome is cut into pieces that adjoin one another or, ideally, overlap, then one can easily determine the "correct" order of the DNA segments in the cell and thereby obtain a coherent picture of the placement of the markers copied out with PCR and their neighboring genes.

Another method for determining the genetic structure begins at the level of the messenger RNA molecules, which can be transcribed with reverse transcriptase into cDNA (complementary DNA). By now, several hundred thousand such partial sequences of expressed genes are known. In hybridization experiments one can now attempt to ascertain the genomic locations at which the associated genes are to be found. This method of sequencing of cDNA clones and obtaining ESTs (expressed sequence tags) makes three things possible: First, it gives knowledge of the structure of a gene transcript and thereby limited information about the structure of the gene itself; second, with the help of a gene library and a cDNA one can determine the location of the gene; and third, the incidence of ESTs gives information about the conditions under which a particular gene is expressed. In this way "transcription maps" can be produced.

## Sequencing

Sequencing genes and eventually genomes is by far the most fundamental but also the most costly way to understand their structure and function. The human genome contains $3 \times 10^9$ base pairs (haploid genome). Some laboratories today are capable of sequencing several million base pairs per year. The Sanger Center, in Oxford, and the group at Washington University in St. Louis together provide almost a million base pairs of new sequences every week, and they will expand their capability within the next few years. Sequencing methods are being continually improved, as are the methods of data storage and of comparing sequence data. Originally, it was planned that all the nucleotides of the human genome would be determined by the end of 2005. Today it looks as if this ambitious goal will be achieved earlier.[35]

It is worth noting the fact that some smaller genomes have already been completely sequenced. The first organism whose complete sequence was published (1995) was *Haemophilus influenzae* (1.8 Mb) and soon thereafter *Mycoplasma genitalium* (0.58 Mb). Since then other organisms, for example *Methanococcus janaschi,* a bacterium that can live at unusually high temperatures, have joined the club. We should also note the recent successful conclusion of the internationally sponsored yeast genome project. Here it was a question of a 15 Mb genome, which was sequenced in just under a year by various European and American laboratories. Other organisms, such as *C. elegans, Drosophila melanogaster*, and *E. coli* (of course), are being studied alongside the human genome (Figure 4.10).[36]

Almost all scientists working on sequencing projects report that about half of the alleged genes (open reading frames) sequenced by them cannot be allocated to known genes. That is, from their structure one can deduce nothing about their possible function. In the sequencing of new sequences one meets essentially four situations:

1. A sequence can allow one to draw conclusions about the biochemical as well as the physiological function of the new gene.
2. A sequence can be associated to a particular biochemical function, say an enzymatic function, without it thereby being possible to establish a unique physiological classification. On chromosome 3 of the yeast genome there were found, for example, five new protein kinases, whose biochemical function is clear: They phosphorylate proteins,

## A Comparison of the Viral, Bacterial, and Human Genomes

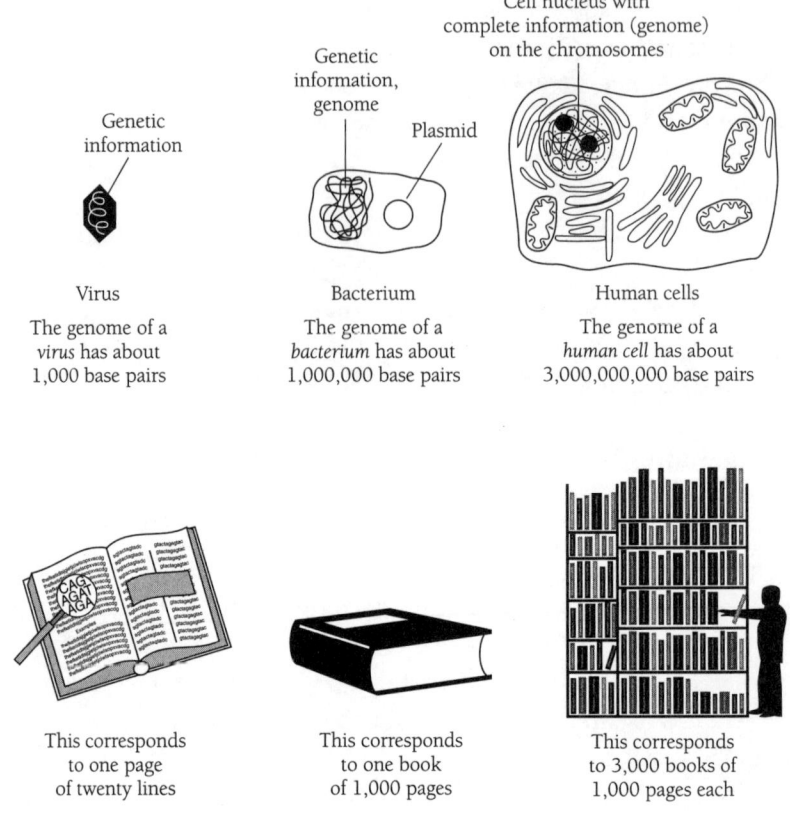

Figure 4.10.   The variety of informational content in various forms of life represented as various quantities of printed information.

whose physiological significance for the yeast cell nonetheless remains mysterious.

3. A sequence is unknown or is correlated with a sequence in other organisms about whose function also no information is available.

4. One finds a sequence that corresponds to that of another organism, but from this correspondence one can draw no conclusions about function. For example, there is an extensive homology between the gene YCL 17c on the yeast chromosome 3 and the NefS protein of nitrogen-fixing bacteria. Although similar gene sequences have been found in *Lactobacillus*, *Bacillus*, and *E. coli*, there are no functional conclusions to be drawn, since neither the last-named bacterium nor yeast is capable of fixing nitrogen.[37]

From this example it is apparent that the association of a particular function with a particular gene is not trivial, even if one believes that one has already found a function for this gene in another organism.[38]

## Understanding the Function of Genes

In the coming years, the various efforts to sequence genes will produce an enormous quantity of data that will have to be interpreted. All these data must be recorded and stored in data banks, which already contain much structural data with previously allocated functions. The comparison of sequences, of sequence homologies, and of functions that can be tied to particular sequences in certain organisms will help to bring this world of data into order. Without electronic data processing, without the new discipline of bioinformatics presently emerging, this flood of data cannot be mastered. Already today many research organizations describe themselves as "gene heavy." With this term they indicate that they possess more genetic information than they are able to digest. The most difficult task of genome research over the next several years will be to figure out the function of the one-third to one-half of genes that exhibit new types of structures that offer no key to determine their function. For this there is a host of strategies, all of which have their weak points and limitations. Some of them should nonetheless be described at this point.

Even when the structure and function of proteins cannot be interpreted, they nevertheless contain certain structural characteristics and domains that also appear in other proteins. Through combinatorial chemistry, compounds of low molecular weight can be produced that bind to such domains. Wherever such binding molecules elicit phenotypic changes in cell cultures, one has then in hand the first key to the possible function of the protein in question. Perhaps the most interesting method of investigating the function of unknown genes and the proteins coded by them comes from developmental biology. Often the cascades of gene activation and deactivation that lead to the development of an organ, for example an eye or a wing in *Drosophila melanogaster*, are today well known. Further genetic and biochemical developmental paths relating to the cardiac and circulatory system and other bodily organs and systems will be understood in detail in the coming years. Genes with unknown functions can be brought into the signal pathway by corresponding organ-specific promoters, which, for example, lead to the development of a compound eye in the fruit fly. When the expression of such a transgene leads to a phenotype, then one can ask whether this phenotype can be brought back to normal by the mutation of other genes. Genes that meet these criteria are potential targets for drug intervention. In this way it is also possible to draw conclusions about the proteins with which the genetic product to be interpreted interacts during the development of particular organs and tissues.[39]

Even so simple an organism as *Saccharomyces cerevisiae* (baker's yeast) can be pressed into service in research into the function of known genes. The yeast genome contains about six thousand four hundred genes. In its entirety these genes represent something like a set of instructions for the structure and operation of a eukaryotic cell. When a yeast cell grows in a very complex medium, it expresses only about one-third of these genes. The remaining two-thirds are used only when the cell has to reproduce or survive under suboptimal conditions. Some research groups are attempting to produce a "minimal" yeast cell, that is, a strain in which all apparent redundant genes are reduced to obtain a minimal set of genes that allows growth in a complex medium. Thus one produces a cell that can grow on complex media but not on simple ones. Through the step-by-step introduction of unknown genes that permit growth on minimal media, one obtains indications about their function.[40]

All these approaches are thus far in their experimental stages. The results obtained to date are few. The most important problem of genome research seems to reside in associating to newly discovered genes, to their regulatory elements as well as other genome components, which today are not at all understood, a physiological or developmental function.

In spite of these difficulties it can hardly be doubted that the rapidly growing knowledge about the structure of genes and genomes will also lead to a better understanding of function. We will learn what role is played by individual genetic products in the organism, what other proteins they interact with, and how they take part in regulatory processes. An understanding of the function of genes united in the genome will also lead to the result that we will get to know many pathophysiological feedback systems that are now obscure. In this way many new possible approaches for drug research, which until now have remained closed to us, will open up. Genome research will make accessible to drug research a large number of new targets for drug therapy. How large this number might be is difficult to predict. The human genome contains, according to current estimates, about 100,000 genes. If ten percent of these genes code for the synthesis of proteins, which represent the possible targets for drug therapy, then there would be about 10,000 proteins. By way of comparison, the entirety of drugs available today influences 483 biochemical targets (including the targets in viruses and microorganisms that can be reached by chemotherapy and antibiotics). With genome research the number of potential pharmacological attack points will be at least an order of magnitude greater.

Great significance must be attached to developmental biology, not only from the viewpoint of molecular biology, which searches for the function of unknown genes, but also from that of drug research. Many of the diseases that plague modern societies, such as arthrosis, connective tissue diseases, diseases of heart and circulatory system, and neurodegenerative diseases like Parkinson's disease and Alzheimer's disease, have ultimately as their basis the loss of specific connective and organ structures: joint cartilage, vascular endothelium, heart muscle, dopaminergic or cholinergic neurons. The removal of these destroyed tissues must be compensated by regenerative processes. These "therapeutic" regenerations to be introduced could follow the same laws by which tissues and organs are constructed during embryonic development. If we know the control processes by which certain genes are turned on and off, and if we could reactivate in a controlled and time-sensitive manner the processes that occur during the development of an organ in order to compensate for defects that have occurred, then we would have found the logical method for treating almost all the diseases of modern society. Genome research and developmental biology will eventually put us in such a position.

These are long-term views. For the coming years the genetic data banks will support research and drug development in a way comparable to the support given by well-equipped and well-ordered libraries. If a drug researcher is on the trail of a particular mechanism, say the development of insulin resistance in tissues, then these data banks can already be of great use today. Let us take a specific case: The insulin receptor is activated by the binding of insulin. This activation takes place biochemically in a series of phosphorylations. It has long been assumed that the deactivation of the receptor must take place by means of a specific phosphatase. Now, there are many phosphatases. If a data bank gave information about the structures, the distribution among organs, and the intracellular localization of phosphatases, then the selection of the enzymes that were possible candidates for an inhibitory medicine would be made much easier. The *immediate* usefulness of genome research and its results for drug therapy consists, then, in an acceleration and simplification of the work process, not yet in a revolutionary change in the course of research. If genome research today makes pharmacological projects easier and accelerates them, then this influence could lead to an increased number of marketable medicines in ten years at the earliest. Well into the first decade of the next century, the combined use of genome research and drug research will not yet elicit a spectacular increase in the frequency of innovations. Thereafter, to be sure, the pharmaceutical innovation process should speed up considerably.

On what are these optimistic assumptions based?

1. The expectation that the number of identified targets for drug therapy will grow drastically in this period.

2. The assumption that combinatorial chemistry will provide so many new compounds that the newly discovered targets can be functionally manipulated even without the primary deployment of rational synthesis.

3. The emerging certainty that the automation and miniaturization of screening techniques will make it possible to bring the multiplicity of biological targets into a productive relationship with the yet greater number of possible chemical compounds. We finally have at our disposal today information systems that make it possible to sort through and compress the data accumulated in such large numbers of experiments so that they will be available for further use.

One more word before we leave this analysis of genome research and drug therapy: According to estimates by experts, five to ten thousand of the hundred thousand human genes code for the synthesis of soluble proteins. If just a few percent, perhaps five percent or at most ten percent, of these soluble proteins are suitable themselves as medicines, then this would open many new possibilities to classical biotechnology as we know it today. Up to now about forty recombinant proteins have been introduced or are about to be introduced. It seems that this is just a small part of what can still be expected.[41]

## Gene Therapy

Oswald Avery and his colleagues were the first to show that the transmission of genetic material from pneumococci of phenotype S (= smooth) to pneumococci that grew in rough colonies led to a transformation in a small percentage of the developing colonies: Colonies that were originally rough took up DNA of the smooth strain and then also grew in smooth colonies. The conclusion that genes determine the form and function of cells and that DNA is the chemical carrier molecule for genetic characteristics led almost inevitably to further attempts to alter the phenotype of cells by means of the transmission of foreign genetic material. At first, such "transformations" were carried out in bacteria. Later, scientists made attempts with eukaryotic cells as well. In the course of time a variety of methods were developed by which DNA can be inserted into mammalian cells: Microinjection, the transfer of genetic material with DEAE-dextran, or adsorption on calcium phosphate and electroporation.[42] In their entirety these experiments, carried out with various types of cells and with a variety of genes, confirmed that one can also transform animal cells and that transformed cells often alter their phenotype. To be sure, the rates of transformation that were obtained with these physico-chemical methods on primary human cells were much too low to be used as a basis for gene therapy. Other, more effective, methods had to be developed; of course, for this, viruses occurring naturally in humans or in animals seemed like good possibilities.[43]

It has long been known that viruses can transmit genetic material from one cell into another. In the early years of their discipline molecular biologists worked almost exclusively with bacteria and bacteriophages. Certain bacteriophages such as λ-phages can lysogenically infect bacterial cells; that is, they incorporate their DNA into the host genome, but for the time being leave the host cell in peace. The phages reproduce only by an external stimulus, such as UV light, certain chemicals, or in the case of temperature-sensitive mutants by raising the temperature: The lysogenic infection becomes lytic. When the phage replicates itself, it often takes host DNA from its site of integration with it and incorporates it in the resulting phage particle. When such phage particles then infect a new bacterial cell lysogenically, the host DNA, which often codes for a specific characteristic, is taken into the new cell. In this way a lysogenically infected bacterium can become, for example, resistant against tetracycline or other antibiotics. With phages that accidentally insert themselves at some place on the bacterial chromosome, one can transfer almost all characteristics of one bacterium onto the other. The transfer of genetic material by phages is called *transduction*.[44]

If bacterial genes were "transduced" by bacterial viruses, maybe one could use mammalian viruses to shuttle genes among mammalian cells. Of greatest interest in this connection were viruses that in analogy to certain DNA phages integrate themselves into human DNA. However, viruses are frequently pathogens: They kill cells off and cause disease symptoms. Thus it was necessary to "reconstruct" natural viruses in such a way that they would still be suitable as a vehicle for DNA while presenting no danger for the infected cell. This at first succeeded quite convincingly with mouse retroviruses. The structure of such viruses is quite simple. They contain one copy of the coding sequences for viral proteins, such as for the nucleus (gag), for the virus polymerase (pol), and for the envelope (env). (See Figure 4.11.) At both ends of the genome they carry long terminal repeats (LTRs), sequences that contain control elements for the expression of the genes, and finally a signal for the "packaging" of the RNA, the ψ (psi) signal. Based on this structure, a method for producing recombinant retroviral vectors could be developed. A "packaging plasmid" contains the provirus genome without the packaging signal. It can produce only empty envelopes, because the RNA without the ψ-signal cannot be incorporated

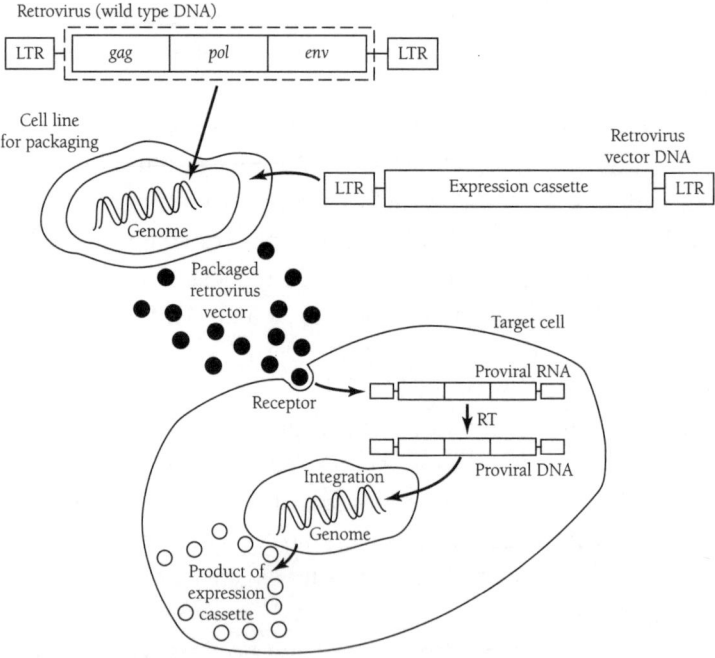

**Production and Transfer of Retrovirus Vectors**

Figure 4.11.   A particular gene is inserted into a retrovirus as an expression cassette in place of the viral genes gag, pol, and env. The expression cassette also contains the packaging signal ψ. A cell line is infected with the transformed virus. At the same time, the cell is infected with the wild-type virus, from which the packaging signal has been deleted. The infected cells now produce a virus that is infectious but no longer capable of reproduction. These viruses can now be used to infect target cells, e.g., lymphocytes. The desired protein is now synthesized in these cells after integration of the retroviral vector.

into a virus particle. The packaging signal is intact in the "vector plasmid"; yet here the viral sequences for gag, pol, and env sequences are replaced by exogenous genes. When one infects cells with both viruses, then through recombination virus particles arise that contain the transcript of an exogenous DNA instead of the viral proteins gag, pol, and env. On the other hand, the packaging plasmids see to it that sufficient viral protein is produced for constructing virus particles with the "exogenous" RNA. The technique was further simplified in that cell varieties were produced that already contained the packaging signal in their cellular genome. Such cells are well suited for the production of large numbers of retroviral vectors.

In such vectors one can put up to nine kilobases, that is, 9,000 nucleotides of exogenous RNA. Their advantages lie in a wide host range, a high rate of transduction, stable integration in the host genome, and the possibility of producing large quantities of the vector. But retroviral vectors also have their drawbacks: They infect only cells that are in the process of dividing. Furthermore, the site of integration in the genome is not specific to the target cell. The provirus can integrate itself everywhere and in this way, at least theoretically, cause genetic impairment. Finally, through recombinatory processes, packaging cells may give rise to reproductive viruses that can cause pathogenic effects.

For these reasons there is a constant search for new vectors. On the basis of adenoviruses, which infect several types of cells and whose infectious activity is not bound to cell division, vectors have been developed that can accept a "maximum permitted load" of up to 7.5 kilobases (Figure 4.12). Here, too, cells have been developed that carry specific genes, such as E1, which is necessary for viral replication, in their chromosome and that produce the E1 protein constitutively. In such cells adenovirus vectors (adenoviruses that carry the gene specific to transfection) can be produced in large yields ($10^9$ copies per milliliter). Advantages of vectors developed on the basis of adenoviruses are a high transduction coefficient, the already mentioned high concentration of virus, and the applicability of these vectors to cells that are not in the process of division. Disadvantages are these: The only transient expression in the target cells; the immunogenicity of the viral proteins, which can cause local inflammation; the cytotoxicity of the particles; and the difficulties in the production of recombinant vectors. A further disadvantage consists in the fact that adenovirus vectors in the cells remain episomal; that is, they remain outside the host genome, and consequently, they quickly become diluted in a population of dividing cells.

## Production and Transfer of Adenovirus Vectors

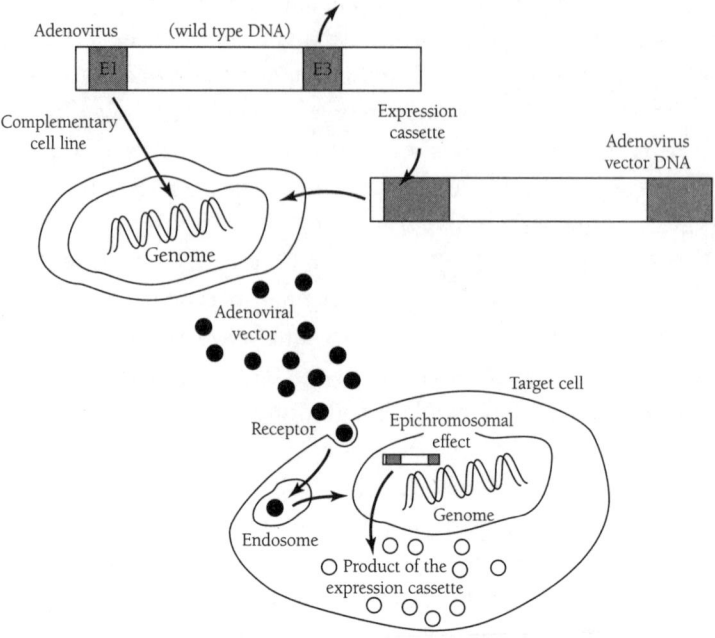

**Figure 4.12.**    Insertion of a gene into an adenoviral vector for gene therapy. The basic processes are analogous to those of Figure 4.11. Here the adenovirus vector is not inserted into the host genome; rather, it remains as an "episomal" structure that can no longer reproduce itself and therefore is thinned out in the cellular population after several divisions.

Work is progressing on the development of many other vectors. Worthy of mention here are the vectors derived from herpes viruses. With them the transfer of genes and their stable integration in neuronal cells could become possible.

AAV (adeno-associated virus) is a nonpathogenic parvovirus that is incapable of independent replication. Vectors derived from AAV integrate stably on human chromosome 19, in a region that apparently contains no sensitive genes.

Of course, further attempts are being made to produce nonviral vectors and to formulate pharmaceutically these structures in such a way that they can be easily taken up by cells. The lipid envelope of such particles contains for the most part synthetic cationic lipids, which surround the negatively charged plasmid. A disadvantage of these liposomal particles is that in large measure they are taken up by cytoplasmic organelles and broken down. It has not been possible till now to construct lipid envelopes that preferentially bind to cellular membranes.[45]

It has already been pointed out that introducing genetic material into human or animal cells and expressing foreign genes in this new environment are possible. Therefore, one prerequisite for gene-therapeutic experiments in animals had been fulfilled. After the most important safety considerations for some of the available vectors could be dealt with, and after the effectiveness of gene-therapeutic techniques had been proven in animal experiments, gene-therapeutic interventions could be tried in human patients. In this, two basic strategies were followed: one in vivo method and one ex vivo. In the ex vivo method, cells were taken from the patient (or experimental animal). These cells were then transfected in vitro, subsequently reproduced, and then returned to the patient. In the in vivo method, which from a purely technical standpoint is closer to drug therapy, the genetic constructs are injected directly into the patient (or experimental animal), either locally or systemically (Table 4.2).

The first protocol for gene therapy in humans aided in the treatment of two children with severe combined immunodeficiency.[46] Children who suffer from this disease carry a defective gene for adenosine deaminase, the key enzyme in the synthesis of purines. Blood lymphocytes were removed from patients with this defect. Subsequently, the cells were stimulated to divide in vitro by stimulation with antigen. The dividing cells were then transfected with a retroviral vector that contained the intact adenosine deaminase gene with the accompanying regulatory elements. In fact, it could be shown that the transfected lymphocytes, which subsequently were reproduced in vitro by means of interleukin-2, correctly expressed the adenosine deaminase gene. The lymphocytes, transfected and reproduced via interleukin-2, were subsequently readministered to the patients. This

Table 4.2.   Basic strategies of gene therapy.

| Category | Action | Purpose |
|---|---|---|
| Ex vivo | Removal of cells from the host<br>Introduction of the foreign gene<br>Reproduction of the cells<br>Injection of the cells | "Drug delivery"<br>Restoration of a missing function<br>Vaccines for the treatment of cancer<br>Cell therapy against tumors |
| In vivo | Direct injection of a genetic<br>  construct into a patient<br>Systemic application or<br>  local application | e.g., Antisense nucleotides<br><br>e.g., Inhalation in cases of cystic<br>  fibrosis |

procedure was repeated a number of times over several months. In the first of the two patients who were treated with this procedure there was indeed a lasting clinical improvement. It is necessary to add, however, that in addition to the "gene therapy," these patients received intravenously a chemically modified adenosine deaminase (with polyethylene glycol). One cannot, therefore, determine with certainty what portion of the clinical improvement is to be allotted to the treatment with adenosine deaminase, and to what extent the gene-therapeutic intervention is to be considered. That the transfection of lymphocytes played an important role in the therapeutic success can nonetheless be concluded from the fact that lymphocytes with the altered phenotype were detectable in the circulating blood for years after their application, though their numbers varied over a wide range. The same treatment can be carried out with stem cells from bone marrow or with progenitor cells from peripheral blood. When stem cells are transformed, the transformed lymphocytes remain detectable for many years. To be sure, in this case, too, the patients were treated, alongside the gene therapy, with pegylated adenosine deaminase.

A strategy that likewise falls into the ex vivo category is employed frequently in oncology. At present, experiments are being carried out mainly on renal adenocarcinoma and melanoma. Tumor cells are taken from patients. Then the cells are reproduced in vitro. Subsequently, they are transfected with a retroviral vector that contains the information for the synthesis of GM-CSF (granulocyte monocyte colony-stimulating factor). After this step they are again replicated for a few generations. Cultures of tumor cells as homogeneous as possible are then irradiated until they can no longer divide but are still capable of protein synthesis and secretion. They are then injected as a vaccine into the patient at intervals of one to two weeks. The results of these procedures cannot yet be evaluated. There have been impressive, but anecdotal, reports about clinical improvements (remissions) and about increases in life span. At the time of this writing, there were not enough data for a statistical evaluation. In spite of some encouraging anecdotal results, there is, strictly speaking, not yet any clinical proof of effectiveness.[47]

On the side of the in vivo method we should mention attempts to introduce the mutant cystic fibrosis gene into adenovirus vectors through inhalation into the bronchial epithelium. Thus far, these attempts have brought no conclusive success. To be sure, it could be established that the cystic fibrosis (CF) gene is taken up in cells of the bronchial system and expressed. The inflammatory reactions that accompany this treatment speak, however, against a broad application of this technique (Figure 4.13).

The American Recombinant DNA Advisory Committee, which his oversight over gene-therapeutic experimentation, had by June 1995 approved 106 gene-therapeutic protocols: Eighty-one of these protocols had a therapeutic purpose; twenty-five fell into the category of marker experiments. These are experiments in which specific cells are marked with an easily recognizable gene, which allows tracking the movement of these

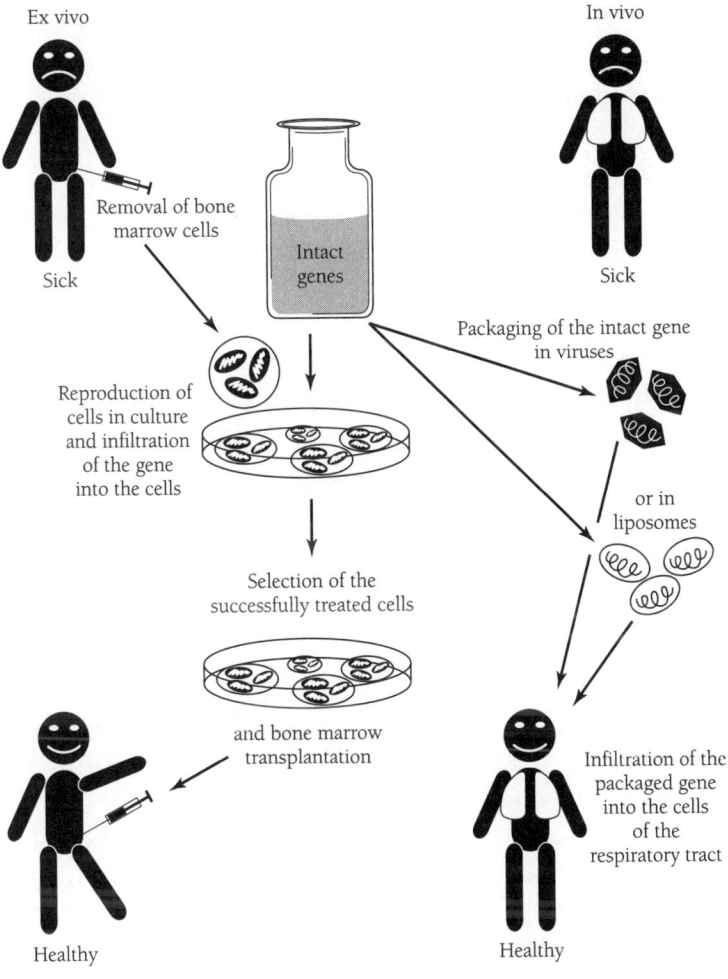

Figure 4.13.    Ex Vivo and In Vivo Gene Therapy. (Akzente 7, *Gentechnik: Eingriff am Menschen*, 3rd Edition, Herbert Utz Scientific Publishers, Munich, 1999.)

cells in the organism. In seven previous studies (considerable adenosine deaminase deficit, familial hypercholesterolemia, cystic fibrosis, and solid tumors), gene-therapeutic intervention was able to obtain desired, though not curative, biological effects.[48]

If we attempt to judge gene therapy as a therapeutic discipline today, we could take the following position: Gene therapy is a method of treatment by which somatic cells of the patient being treated are transformed by genes introduced from an external source. Although the fundamental feasibility of this method may be considered proven, the development of the relevant techniques is still in its infancy. In specific cases, clinical results have been obtained that have been celebrated as therapeutic successes. Such results are encouraging, but up to now they remain anecdotal. Were we to demand the same level of proof of safety and effectiveness that we apply to drugs, we would then classify gene therapy as neither safe nor effective.

Through uncritical coverage in the lay press, unrealistic expectations of immediate results have been awakened in the public. Unfortunately, many scientists and physicians also have fallen victim to the wishful thinking that has surrounded the modest kernel of well-established facts like a colorful, noisy crowd of onlookers surrounding some small object of interest. If we concretely inquire into the difficulties that must be overcome in order that gene therapy become a technique that can be set alongside drug therapy, we must consider the following points:

- All human studies have been marked by an unexplained degree of inconsistent results. For example, the proportion of transformed lymphocytes in the case of the above-mentioned children with severe combined immunodeficiency ranged between 0.1 and 60%! In the experiments on affecting cystic fibrosis through the inhalation of adenoviruses with the intact transporter gene, at most five percent of the bronchial cells were successfully transformed. In most experiments the transformation rate was considerably less. And in most of the marking experiments with human bone marrow cells only a temporary successful transfer of genes was obtained.[49]

- Apart from a few exceptions, the standardized production of large quantities of vectors still presents considerable difficulties. An industrial application of gene therapy must be built on techniques that are reproducible and standardizable.

- Particularly in the case of in vivo gene therapy one must place demands on the quality of the vectors used and on their pharmaceutical formulation, demands that today have not been met. The injection of a specific gene-therapeutic preparation must make it possible to insert particular genetic constructs exclusively or almost exclusively into particular cell populations or organs. For this, special pharmaceutical preparations are required. Theoretically, this would be a question of surface receptors as specific ports of entry. The rates of transfection achieved must be high enough in individual cases to accomplish a local or systemic effect. For this the transfection rates of a few percent observed to date are inadequate. Further, it must be required that the genetic constructs introduced into the cells either be stably integrated into the genome without causing injury or be able to exist and function episomally in stable cell populations for extended periods of time. For this the administered gene dosage must lie within clinically determinable dose–effect relationships.

None of these requirements seems unreachable in principle. On the contrary: With the AAV virus we already have at our disposal a vector that integrates itself at a particular place in the genome. Even so, it seems possible to develop formulations that can be addressed selectively to particular cells or organs through the use of receptor specificities. But it will be years, if not decades, before these techniques have matured to the point where they can be used generally and with relative safety, which is the case today with large segments of drug therapy. After overcoming these major technical difficulties, gene therapy could certainly develop rapidly. We are learning continually about new genes and their functions, and from this knowledge we can develop new ideas about the best forms of gene-therapeutic intervention. Many of these ideas can be verified with animal experiments. Once the technical problems that have been described above have been solved, gene therapy can develop rapidly.

If all current gene-therapeutic experiments lead to successes, large portions of drug therapy would become obsolete. Many diseases could be controlled on a long-term basis by a single gene-therapeutic treatment or a series of treatments occurring over long intervals. However, it seems a long way to that point. It seems to the author that over the next ten to fifteen years the merging of results from genome research with new techniques of chemistry and automated screening methods offers more convincing prospects for therapeutic success than does gene therapy.

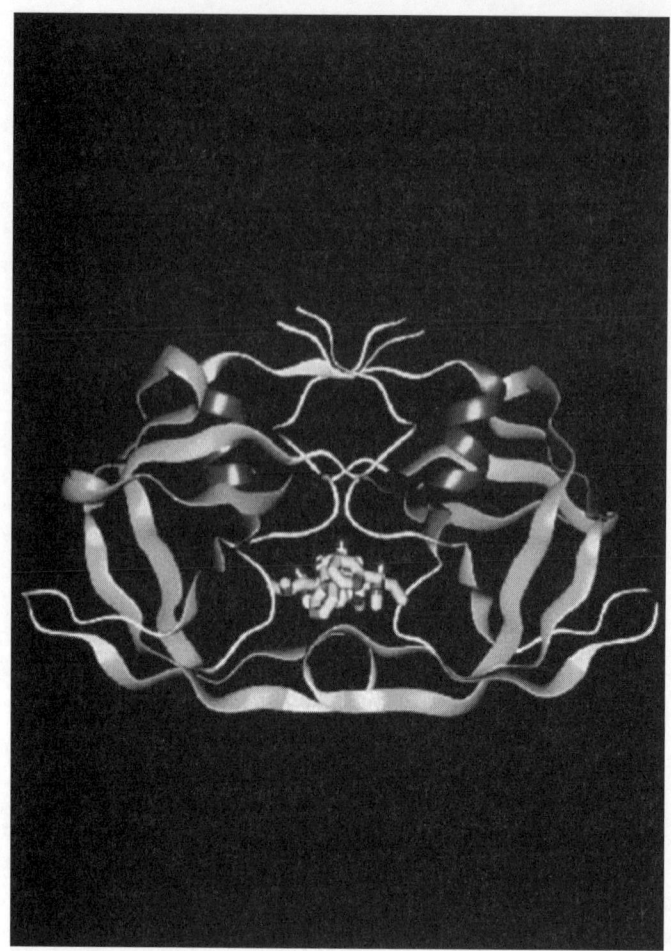

Figure 5.1.   Model of a protease of the HIV virus with a protease inhibitor in the catalytic center. HIV protease is a symmetric molecule composed of α-helical components and β-sheets. The two symmetric halves of the protease are shown here in red and blue. The inhibitor (yellow) is exactly at the location that normally is used for the cleavage of the HIV precursor protein into functional virus proteins. One can see that the inhibitor almost completely fills the free space, the "pocket," between the two halves of the molecule. (From Flexner, C., "Drug Therapy: HIV-Protease Inhibitors," *The New England Journal of Medicine,* Vol. 338, pp. 1281–1292, Copyright © 1998, Massachusetts Medical Society. All rights reserved.)

# 5
## ∧∧∧∧

# A Medicine Is Born:
# Research,
# Development,
# and Registration

Now that we have examined the scientific and technological traditions that made modern drug research possible, we should pause at this juncture and ask, How are drugs discovered *today*? Drugs must first be synthesized as molecules or discovered as "biological activities." Then their biological and chemical properties must be precisely described. Such a characterization should also include the formulation of preliminary hypotheses concerning the possible clinical uses of the newly found substance. All activities in this sphere are comprehensively listed under the heading "research." The development process that then follows comprises the production of a substance in quantity; its processing into the form of a drug, for example, into tablets or capsules; and its subsequent clinical testing, including ascertaining any side effects. When all these individual steps have been accomplished and reliably documented, then the new substance can be submitted to the appropriate governmental drug authority for registration. In this chapter we shall outline these various activities, beginning with research. What are the considerations and questions and what are the experimental approaches that lead to the discovery of new drugs? What, in short, constitutes the research out of which we expect the discovery of new medicines to emerge?

There is no single answer to these questions because the traditions at which we have been looking are even today blended in various ways. In one case a biological test is the starting point in the search for a new drug, while another time it is a chemical compound or a class of compounds, and in yet a third case it is a biochemical mechanism.

## Research Strategies

A person given the opportunity to observe the research that goes on at the large pharmaceutical firms today would conclude that most treatment-oriented research projects start with an understanding of biochemical mechanisms. Such mechanisms can be represented by physiological processes that may be pathologically altered in a diseased organism; alternatively, they could be identical to biochemical reactions in a cell that the therapist would like to eliminate or to prevent from  proliferating. This could be a bacterial cell, a parasite, or perhaps a cancer cell.

Let us first consider an example from the first of these categories. For decades now, basic research in physiology has concerned itself with humoral and neural mechanisms that regulate blood pressure. Such research also takes place in industrial laboratories. We have already spoken of adrenergic mechanisms and their pharmacological modulation by beta-blockers. But the release of epinephrine and norepinephrine represents just *one* regulatory circuit. Yet another substance, one that reduces the diameter of small arteries (precapillary arterioles) and thereby raises the resistance to flow, is angiotensin II. It has several loci of action, of which two are particularly important: One is its effect in narrowing the small arteries, particularly in the region of the abdominal viscera, in the splanchnic area, and also in the region of the kidneys. Secondly, it leads to an increased formation of aldosterone in the glomerular zone of the adrenal cortex. Aldosterone, in turn, causes an increased reabsorption of sodium in the distal renal tubules. The increase in sodium concentration is followed by an expansion of the volume of extracellular fluid. Both mechanisms lead to an increase in blood pressure. After it was learned that angiotensin II is a peptide that causes a sharp increase in blood pressure, it was postulated that inhibiting the action of this substance must lead to a lowering of blood pressure in a way that might be of therapeutic value. Two pathways suggested themselves: First, one could attempt to block or weaken the effect of angiotensin by preventing the binding of this peptide on its receptor. Angiotensin II (as well as angiotensin III, a heptapeptide) produces essentially all of its pharmacological effects via the $AT_1$ receptor. In order to achieve this, pharmacological tests in which a biological effect could be measured as a function of the occupation of the angiotensin receptor ($AT_1$) by its agonists had to be developed. This proved to be a long and tortuous path. It led to peptide analogues such as saralasin and eventually to the imidazole-5 acetic acid derivatives, among which two substances were found that were very weak nonpeptide angiotensin II receptor antagonists but that nevertheless acted very selectively: They no longer possessed further partial agonistic activity. From these weak antagonists a group at Dupont then developed Losartan (DuP753). This substance was, in turn, the prototype for a further fifteen substances that are already on the market or in clinical testing.[1]

The second strategy, which bore fruit earlier, began with the knowledge that active angiotensin II is released from a precursor protein, angiotensinogen, which is produced in the liver. First, angiotensin I (a decapeptide) is produced under the influence of the enzyme renin. In a further proteolytic step a dipeptide (His–Leu–COOH) is split off the carboxyl terminal end by the angiotensin converting enzyme (ACE). This angiotensin II has, like angiotensin III, which arises from the splitting of an additional amino acid at the N terminal end by an aminopeptidase, a great effect on blood pressure. Injected into the bloodstream, it is forty times as effective as norepinephrine.

If it were possible, so some pharmacologists argued, to inhibit the release of the effective agent from its precursor, angiotensin I, this could represent a strategy for treating increased blood pressure. Therefore, one would have to study the enzyme responsible for converting angiotensin I into angiotensin II. In the pursuit of this strategy, scientists found that the activating step, which is identical to the cleavage of two carboxy terminal amino acids, is very specific. Angiotensin converting enzyme (ACE) recognizes the correct sequence at the carboxy terminal end of the precursor peptide (–Pro–Phe–His–Leu–COOH) and transforms inactive angiotensin I into active angiotensin II by the splitting off of the dipeptide His–Leu. The work of pharmacologists and chemists now consisted in offering the enzyme a substrate that was chemically similar to the part of the peptide that the enzyme recognized and cleaved. Researchers were favored with the circumstance that in the sixties, peptides had been found in the venom of the pit viper that inhibit ACE and act as weak reducers of blood pressure. Knowledge of these peptides was the starting point for the synthesis of teprotid, a nonapeptide, which on the one hand inhibited angiotensin convertase and on the other reduced blood pressure in human clinical trials.[2] First, shorter peptides for the purpose of deceiving the enzyme were synthesized. Later, when this had been accomplished, Ondetti and Cushman, together with their collaborators, were able to produce compounds that accomplished the same goal, although they were no longer peptides.[3] Such compounds are called peptidomimetics. Thus the first tasks consisted in the isolation of an enzyme and the determination of its activity. Then it was necessary to understand how ACE recognizes its substrate and cleaves it. And finally, and this was a job for the chemists, "substrate traps" had to be manufactured that could trick the enzyme. Enzymologists, protein chemists, and synthetic chemists had to—and must in all similar cases—work closely together to accomplish this first goal. Afterwards, or parallel to these basis activities, the "best" enzyme inhibitors had to be tested for their pharmacological effects. That is, it was necessary to verify that the inhibitors that were found worked reliably and effectively in various animal species. In such pharmacological experiments one must determine how the substance will be administered. Can it be given orally, that is, in the feed, or must it be administered parenterally, that is, by injection? What is the relationship between dosage and pharmacological effect? How long does it take to achieve effective concentrations in the tissues with a given dosage? How is the substance metabolized and excreted? Can the pharmacological effect be maintained over long periods of time through regular doses, or is it reduced as a consequence of compensating reactions in the organism? These are just some of the important questions that must be answered once effective inhibitors of a pathophysiologically important enzyme have been found. We have been considering an example in which the inhibition of a pathophysiologically relevant enzyme represented the first pharmacological and chemical goal. In such a case the chemical structure of the enzyme substrate presents the chemist with ideas for the possible structure of the inhibitor to be synthesized. Possibly, he will also make use of the mechanism of the chemical reaction catalyzed by the enzyme to trick the enzyme. He can try to offer the enzyme a false substrate that is recognized by the enzyme and is converted to an intermediate product that now by virtue of its chemical makeup can no longer be transformed into the final product, but the enzyme can no longer separate itself from the false intermediate product. It is blocked. One speaks in such a case of a "transition state analogue." The inhibition of pathophysiologically important enzymes was until now the most successful strategy of drug research. The angiotensin convertase inhibitors, the inhibitors of cholesterol biosynthesis, the above-mentioned carboanhydrase inhibitor, the inhibitors of digestive enzymes (Xenical®, a lipase inhibitor for reduction of fat resorption), and many other examples all testify to the validity of this statement. In chemotherapy as well, enzyme inhibitors play a dominant role: The penicillins and cephalosporins inhibit enzymes for the synthesis of the cell wall, and there are many compounds that inhibit various steps of the synthesis of nucleic acids, most of which are employed in cancer chemotherapy. We might mention here the dihydrofolate reductase inhibitors as well as the inhibitors of inosine monophosphate dehydrogenase, thymidylate synthetase, and many others as well.

If the inhibition of a functionally relevant enzyme is the goal, then a further, but not unlimited, set of chemical structures presents itself to biologists and chemists. A similar situation obtains in the search for receptor antagonists, even if not to the same extent. If one knows the structure of ligands, then one has grounds for the possible structure of a receptor antagonist. To be sure, it is more difficult if the ligand is a peptide or even a protein. In such a case crystals of the complexes of receptors and ligands are needed, which can be studied by x-ray crystallography. Only with such expensive structural analyses does one obtain a glimpse into the spatial relationships in which ligand and receptor meet. Sometimes, useful conclusions can be derived in this way regarding the synthesis of potential receptor antagonists, or even agonists.

## Strategies in the Search for New Active Agents: "Semirational" Procedures Versus "Blind" Screening

We might designate the typical procedure of biologists and chemists in the search for enzyme inhibitors and receptor antagonists—and these are still the dominant strategies of drug research—as "semirational" methods. The structures of substrates and ligands are known, and from this information certain conclusions regarding the possible structure of enzyme inhibitors or receptor antagonists can be drawn. However, in implementing these strategies one is governed by the law of trial and error. The chemist frequently has empirically obtained ideas—though they are in part also speculative—as to what kinds of chemical modifications might produce the desired effects. These ideas are put to the test in biological experiments, for example in an enzyme assay or a receptor–ligand binding assay. The process is semirational, empirical, and logical. However, the results obtained are only partially generalizable and applicable to other situations. It is, then, at least in part a matter of working on a case-by-case basis. In fact, there are only two alternatives to the semirational working method, which even today is still the predominant approach: first, "blind screening" and second, a fully rational modus operandi where the structure of a drug is designed on the basis of a complete set of structural data derived from the target molecule. In the 1970s a number of medicinal chemists were of the opinion that "rational synthesis" was the method of the future. Scientists were confident that they could design pharmacologically active molecules as if on a drawing board, followed by made-to-measure synthesis. Despite considerable progress in our understanding of the structure of macromolecules, this prophecy has yet to be fulfilled. The parameters that must be recorded in each case and then kept track of in the process of synthesis are apparently too numerous and too complex to be computed easily and fully. Only once in his career has the author experienced a case in which the discovery of a new inhibitor, a thrombin inhibitor, was characterized *decisively* by a rational understanding of the molecule's structure. Even in this case the molecule was not developed on the drawing board or the computer screen, though something close to this was achieved.

The other extreme, the disavowal of rationality, so to speak, resides in blind screening. This term describes the random testing of a great variety of dissimilar compounds in one or more biological tests. Such blind screening has been used primarily in antimicrobial chemotherapy.[4] Ironically, here and elsewhere it has achieved little or nothing. Antibiotics have come into existence through adherence to the principle of "antibiosis" enunciated by Paul Vuillemin and through the semirational semisynthetic modification of other antibiotics. Even the sulfonamides were not the result of blind screening. The chemical work that led to prontosil began with chrysoidin, that is, with a dye. Prontosil was the result of serendipity: An active principle was found that had nothing to do with the nature of the molecule under investigation as a dye, even though that had been precisely the reason for studying it. But blind screening? Blind screening knows no theories and no hypotheses, while in the discovery of practically all important active substances it is hypotheses and theories that were at work.[5]

In other areas of drug therapy as well, blind screening has at best led to a "lead substance" that could then be derivatized in the usual semirational way. Modern drug research represents a remarkable mixture of work that proceeds deductively at first and then continues through a process of selection, improvement, and renewed selection that operates according to the laws of trial and error. Blind screening in its true sense has always been only one component of the methodological spectrum and—particularly in chemotherapy—not a very successful strategy at that.

It is worth noting that a not very successful method is returning with a vengeance under the aegis of combinatorial chemistry. And as nearly as we can assess this new development, we must assume that there is justification for it. With the methods of combinatorial chemistry "substance libraries" can be assembled that encompass not just thousands of substances, but hundreds of thousands or even more. And by the correct selection of building blocks to be used in combination, one can also assemble libraries in which the individual substances are not only members of a large "extended family," but belong to a number of different "structural families." In this context one speaks of covering large "areas of diversity." Such "diverse" libraries can in fact be put to use wherever our knowledge of the chemical structure and function of a biological target molecule is meager, with the result that chemists have no satisfactory point of attack, or even none at all. But each "hit" provides a basis for the accustomed—and in experience quite successful—semirational procedure, which consists in the optimization of candidate substances (Table 5.1).

We have mentioned that the optimization of candidate substances proceeds in one or more biological tests that are dictated on the one hand by empiricism, on the other by rational considerations (hypotheses). Certain varieties of combinatorial chemistry, called parallel synthesis, can be helpful in this regard. The principle behind this method consists in modifications of a basic structure designed with the aid of computer software and—when desired—synthesized with the help of automated machinery

**Table 5.1.**   Screening is characterized today to a considerable extent by combinatorial chemistry and automation.

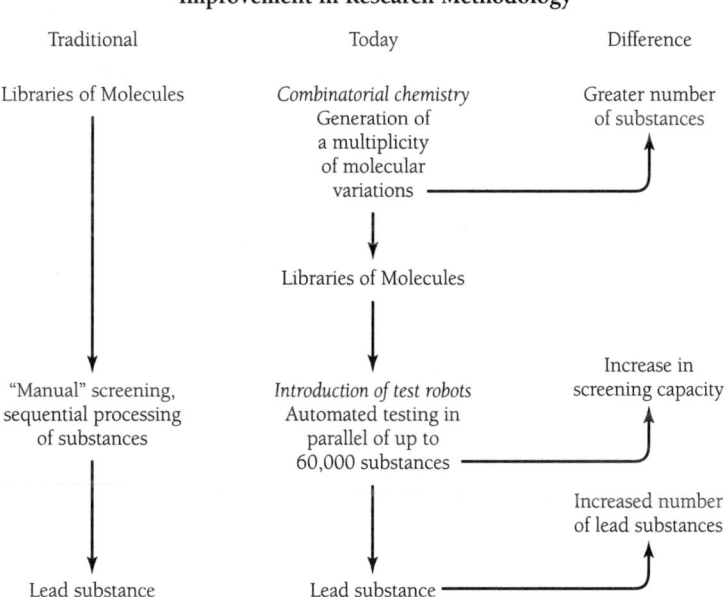

**Improvement in Research Methodology**

| Traditional | Today | Difference |
|---|---|---|
| Libraries of Molecules | *Combinatorial chemistry* Generation of a multiplicity of molecular variations | Greater number of substances |
|  | Libraries of Molecules |  |
| "Manual" screening, sequential processing of substances | *Introduction of test robots* Automated testing in parallel of up to 60,000 substances | Increase in screening capacity |
| Lead substance | Lead substance | Increased number of lead substances |

(robots). The substance can then be automatically subjected to biological testing. Measurements taken from these experiments can then be sent back to the computer, which then on its part bases further syntheses on these results. In this way the computer correlates the "improvement" or "worsening" of the results with changes in physiochemical criteria, for example the lipopophilicity or electrical charge. Sometimes—not always—striking improvements in effect can be achieved in twenty or thirty steps.[6]

## The Search for "Effective" Proteins

In the search for proteins with medical uses, the first step often consists in the identification of all proteins transmitted by a cell to the outside. These proteins carry a terminal $NH_2$ sequence that is recognized and split off in the process of secretion of membrane-bound enzymes (proteases). The "leader sequence" is, so to speak, the passport with whose help a protein is able to leave the cell in which it was created. Genes that code for such proteins can be identified in gene data banks with the aid of appropriate search programs. However, for this one needs first of all very extensive data banks, and second, search programs with which all corresponding genes can be recognized. One can also identify such genes experimentally. The method in question will be called the "gene trap" method. It consists in transfecting invertase-negative yeast cells, which therefore cannot split sucrose, with a plasmid bearing an invertase gene. Since the enzyme coded by this gene has no leader sequence, it must remain in the cell. Such yeast cells are also unable to grow on a sucrose-containing medium. This plasmid is suitable for testing cDNA libraries. If one puts the cDNA to be tested into the plasmid in such a way that the invertase is fused at a particular location with the gene that has been newly taken in, then the subsequently synthesized fusion protein will be secreted: On an appropriate medium, sucrose can now be split into glucose and fructose; and the yeast cells that can utilize these cleavage products grow into a colony. The corresponding gene can then be sequenced and expressed in great quantity. It is now a matter of searching from among the many secreted proteins that can be found in this way for those that have therapeutic value.[7]

These are all current methods for finding new medicinal substances. But the discovery of new candidate substances as well as their optimization in a model that encompasses one or more measures of effectiveness—as we have seen, for example, in the case of computer-aided parallel synthesis—can be automated. But this automation of processes, the identification of molecular targets and active molecules, can be of help only if a scientist is in a position to place the data that are acquired into a meaningful functional context. Eventually, we have to treat diseases. Therefore, we must know how these diseases arise, what functional processes are disturbed, how genes contribute to these disturbances, and what corrective measures should be brought to bear so that a "normal" phenotype, or one approaching normal, might be restored. Therefore, as in the past, we need physiologically or pathophysiologically trained researchers who can provide us with these pictures. The data produced by molecular genetics and genome research have to be brought into a physiological context and must be described in biochemical categories. Only then will they lead to an enlargement of our medical worldview, which is necessary if we are to discover and develop new therapeutic interventions of specific effectiveness. At present, this research—it should here be designated as "molecular physiology" or "molecular pathophysiology"—is being neglected. At the moment everyone is fired up about the fascinating possibilities of genome research, by the potential of combinatorial chemistry, and by the idea of "high-throughput screening." This is true especially of scientists who are far from medicine, and even more so for businesspeople, who may be able to imagine "processes" in the broadest sense, but to whom the complexity of medicine and biology is extremely distant and who may even feel uneasy about it. Here lies a danger for pharmaceutical research that is not limited to the large pharmaceutical firms but that applies to all institutions that seriously take part in drug research today.

The universities must begin again to educate physiologists—physiologists, that is, who have internalized the philosophy and fundamental approach of genetics and molecular biology and who can then go on to redefine and reestablish physiology and pathophysiology on a new, epistemologically more elevated, plane. Only when this occurs will the gain in new knowledge brought about through genomics and through advances in chemistry and screening techniques enable drug research to find and develop novel therapeutics.

Up to now we have spoken almost exclusively of research. We have traced the long path of drugs from the dawn of history into the full light of the present day. We have attempted to understand how drugs were sought in the past, how these searches gradually became systematized, and how the methodology of drug research constantly changed and even today continues to change—to the extent that our scientific understanding of the world grows and we thereby come into possession of new search strategies and new technological instruments. But drugs have not only to be invented or discovered. They must also be developed. Innovations, and especially innovations in drugs, about which we have already spoken in Chapter 1, have two components: first, the creation of something new, and then, the preservation of the new through a rapid development that takes into account all aspects of the drug's effects. The new medicine, which deals with therapeutic needs and with which one can perhaps treat illnesses that previously resisted all treatment, is always the result of research *and* development. Both are important. One function presupposes the other, and they must be seen and evaluated together.

What does "development" mean in this connection? What do we mean when we speak of drug development, and where are its beginnings to be found?

## What Is Development?

Let us begin with a definition: A scientific team has established the validity of a specific biological or chemical idea and has found a new active substance whose structure they know and whose pharmacological effects they can describe. These scientists have made a discovery (or invention). They hold in their hands a new substance. They know the basic chemical and biological properties of the new compound, and they have an idea—at first theoretical, perhaps also already supported by animal experiments—of its therapeutic application. Let us suppose that all chemical and biological results show the substance to be new and highly promising. Our team of researchers is still quite far from having a medicine that can be manufactured in consistent quality and prescribed by a physician in established dosages. To pass from a new-found substance to a medicine, many questions must be answered. Here we shall mention a few of the most important: The pharmacological effects of the new substance must be tested on healthy experimental human subjects and described as they relate to the target population in question. An effective, physician-recommended dosage must be found. Side effects, both common and rare, must be recorded and evaluated from a large number of patients (several thousand). The interaction of the new substance with other drugs must be described, as well as the influence of food intake on its effects. For the clinical trials and for the eventual clinical use galenical forms must be developed and analyzed. The details of production of the medicine have to be elaborated so that eventually a precisely specified synthesis process becomes available. This process must always lead to the same product within very narrow limits. All byproducts must be qualitatively and quantitatively recorded. They should make up, incidentally, only a small and constant portion of the product (in general, less than five percent). Under the rubric of development will be gathered the entirety of operative individual steps that must be completed in order that a new substance— whose structure we know, whose pharmacological properties are likewise known, and about whose clinical application established ideas already exist—can lay claim to being a freely available medicine. In greater detail, there are clinical and nonclinical methods.[8]

## Preclinical Development

In the nonclinical, or preclinical, development stage one differentiates among chemical, biochemical, analytic, galenical, and biological methods. A substance that one wishes to "develop" must first of all be able to be produced in large quantities. For this a new synthesis procedure must usually be developed, since the way the substance was originally synthesized in the laboratory frequently is unsuitable for producing the quantities required in development. It must also be firmly established that the quality and purity of the end product remain consistent (greater than 95% purity, consistency of byproducts). To be able to control the attainment of these goals, analytic tests must be available not only for development, but also subsequently for validation. By "validation" we mean the verification that the methods put into place indeed accomplish what is expected of them.

In the discovery phase of a drug the primary interest is in how the new substance acts on living tissue and on a living organism. During the developmental phase it must also be learned how the organism alters the substance, that is, into what compounds the substance is metabolized and how the original substance and its catabolic products are distributed in the organism, how long they are held in the various regions ("compartments") of the body, and by what pathways and by what mechanisms they again leave the body. For this complex ADME (absorption, distribution, metabolism, and excretion) studies one also needs suitable biochemical verification procedures. So much for an overview of the chemical part of the early developmental stages. For the late clinical phase (phase III) and for commercial production it is frequently necessary once again to develop and validate a new chemical procedure. This should be started as early as possible, that is, already during the clinical phase II, and in certain cases even earlier. At this point one has preliminary information about the side effects of a substance in human beings and perhaps already a preliminary impression of its effectiveness. To be sure, these first clinical data on effectiveness and safety will not be based on a large sample but at most on several hundred patients. To invest millions in a new production process and often in new facilities as well under such conditions requires both the capacity for precise critical analysis and the readiness to take risks.

We come now to biological preclinical development. We have already spoken about ADME studies on animals, usually on two species, of which one should be some sort of rodent. For a long time it was considered necessary to carry out ADME studies on rats and dogs, or even on small primates such as marmosets. Yet these experiments were often disappointing in view of their lack of carryover to human beings. Only in recent years have models been developed from comparative analysis of a variety of animal species that allow a more precise prediction about effects in man. Despite any existing uncertainties, ADME studies on human subjects remain the basis for establishing correct dosages for patients and for the development of appropriate dosage schemes.

Before a new substance can be tested extensively on animals, provisional methods of administration must be developed. These can be orally absorbable forms or injectable preparations. Although the same high demands are not placed on such pharmacological preparations as will later be placed on the formulations for clinical experiments or those to be marketed, they nonetheless require a great deal of often difficult work. Often, problems of solubility must be overcome. The particle size of a preparation must be set, and additives must be selected and then investigated with a view to their biological influences. Even in "simple" cases a high degree of reproducibility and experimental consistency is demanded, to a degree that in a purely research laboratory would almost never be required and certainly seldom achieved. And this canon of reflection, formulation, analysis, stability tests, and new analysis repeats itself once again, in most cases, in fact, twice more: first in the formulation of the first clinical dosage forms and then in the development of the final formulations intended for phase III and marketing.[9]

## How Dangerous Is a New Active Substance: The Question of Toxicity

An important part of the nonclinical development phase is dedicated to establishing the toxic effects of a new substance. The pharmacologist characterizing a new substance must know within what dosage range therapeutic effects appear. In interpreting his findings he must also discover in what dosage his substance produces toxic effects in acute (single) administration or with repeated administration within a twenty-four-hour period. Ideally, this toxic dosage will be far above the pharmacologically necessary dosage. Earlier, the $LD_{50}$ (for *lethal dosage*)—the dosage that in the course of at most two weeks after a one-time administration proves fatal for fifty percent of the experimental animals—was an important measurement for orientation. Today, this measure has lost much of its importance. More significant than the $LD_{50}$ are the relationship between dosage and toxic effect, the temporal course of toxic effects, the determination of those organs or organ systems first affected, and the reversibility of toxic phenomena. In general, such acute toxicity experiments carried out on two species have only an orienting character. In unusual cases, with an appropriately large number of animals and careful analysis they can serve as a basis for the administration of single doses to human subjects for pharmacokinetic purposes.

Subacute or subchronic toxicity tests also permit the researcher to study the toxic properties of a substance over a longer period of time. Depending on the planned clinical application, the period of such studies will be in the range of fourteen to ninety days. In subacute studies the increase (or decline) in the appearance of toxicity effects during continuous application of the substance can be evaluated. To be sure, the length of these experiments, which likewise are carried out on two species (mostly rats and dogs), is hardly sufficient to reveal all the secondary toxic effects of a substance, which perhaps appear only by months-long or even life-long use. In particular, such studies are unsuitable for evaluating the incidence of cancer. In subacute toxicity studies a substance is administered several times daily in three different dosages. However, the lowest dosage should be at such a level that a "no toxic effect level" can be defined. Of course, in all subacute toxicity experiments control groups are employed.

Chronic toxicity studies, which last at least six months—more usually, however, for a year—should answer the following questions:

1. What risks are entailed by long-term use of the new substance?
2. What toxic effects are observed?
3. In what range of dosages do no toxic effects appear?
4. In what range of dosages are exclusively pharmacological effects observed?

5. What is the DTM (*dosis tolerata maxima*), that is, the maximum dosage with no side effects?

6. What is the largest dosage that can be tolerated (without significant lowering of life expectancy)?

Regulatory agencies generally require of the manufacturing firm that these studies be carried out on a rodent species (mouse or rat) and on a nonrodent species (dog or primate).

Carcinogenicity studies are generally carried out separately from the long-term toxicity experiments. Usually, these studies are carried out on rodents (mice and rats) over their entire two-year life span. There are various grounds for determining the choice of dosages. First, one can be guided by the pharmacodynamically effective dosage. One can also use the appearance of toxic effects as a guide. Thus, for example, one could choose the DTM as the highest dosage and then the others lower by a factor of three; that is, the middle dosage would be one-third of the DTM, and the lowest one-third of that. The dosage (or supply of active ingredient) should be chosen, if possible, such that pharmacokinetic endpoints are taken into account, for instance, the maintenance of a minimal concentration in the blood throughout the entire experiment.

Special toxicity studies are directed at local side effects of new compounds. The identification of allergic effects is also part of this repertoire.

After the thalidomide affair, special attention was given to toxicity studies concerned with reproduction. Typically, such studies are carried out in three segments. Segment 1 deals with the possible effects of a new compound on fertility and reproductive behavior in rats; segment 2 studies are designed to detect teratogenic effects in rats and rabbits; segment 3 examines postnatal development.

Today, an entire palette of so-called genotoxicity tests for studying mutagenic effects is available. The purpose of all these tests is to measure the influence of test molecules on genetic material. They are used regularly, and they also contribute to making the risks of carcinogenicity visible as early as possible, indeed, at a point in time when there may be no hard data whatsoever available on the actual carcinogenicity of a substance. The positive outcome of a bacterial mutagenesis test (Ames test) or a micronucleus test can at least be a reason for excluding children and women of childbearing age from the clinical testing of a substance. It can also lead to the result that development is abandoned entirely. It would be pointless to develop a sleep-inducing agent or an analgesic with mutagenic effects. On the other hand, there is no reason that mutagenicity should represent an insuperable barrier to the development of an anticancer drug or a highly effective medicine against a life-threatening infection such as AIDS.[10]

Since 1979, all preclinical studies in the United States whose results become part of the documentation on the basis of which a medicine is to be approved have been subject to GLP (good laboratory practice) regulations. These regulations cover seven distinct areas:

1. Organization and personnel in the research facility
2. Quality of the experimental facility (for example, that of a toxicology laboratory)
3. The operation of the facility
4. Substances tested
5. Protocol and implementation of the experiments
6. Documentation and reporting
7. Adequacy of the laboratory equipment

A study director and quality assurance unit ensure the oversight of all seven areas in the preclinical testing of substances. Legislation requires that "standard operational procedures" be followed by all research facilities. These methodological instructions and descriptions of basic procedures are meant to ensure that minimal standards are adhered to in the planning and execution of experiments. The quality assurance unit (QAU) is charged with seeing to it that standard procedures are followed in all experiments and by all workers in the research facility. The Food and Drug Administration inspects preclinical research facilities throughout the world to oversee the observance of the GLP regulations. Statistics from the first few years after implementation of the GLP regulations indicate an improvement in general standards. At first it was feared that these regulations would lead to an increased bureaucratization of preclinical development. In the long term, it appears that rather the opposite has occurred. Because of the general improvement in quality brought about by the GLP regulations, the FDA and other drug regulatory agencies have been able to rely less on the observance of bureaucratic formalisms.

## Clinical Testing: Phase I

The goal of all clinical testing is to establish safety and effectiveness in human beings of the substance under investigation. According to their scope and structure we distinguish three groups of clinical studies: phase I, phase II, and phase III. They follow one after the other and give the entire development process its structure with respect to time and content.

During the course of phase I, the safety of a new compound is of primary concern. Phase I studies comprise in general between twenty and eighty patients or healthy subjects, and in exceptional cases a larger number of individuals. These are always clinical–pharmacological studies. That is, the medicine being tested is first administered in very small single doses. All the subjects are observed carefully, since at this point there is no information about the safety of the new substance in man. If the single doses produce no ill effects, then the dosage is raised, usually in small steps, until finally a dosage is reached at which acute side effects regularly occur. Such experiments are usually carried out in healthy subjects, who are treated and observed on an inpatient basis. It is very important to record not only the dosage but also the concentrations of the substance in the blood plasma at which the first side effects appear. Once this threshold has been ascertained with some degree of certainty, the new medicine is tested for side effects when administered in multiple doses. In this case as well it is important to determine a dosage range, but also the range of concentrations, within which side effects can be kept at an acceptable level. From animal experiments information about the relationship between plasma levels and desired (therapeutic) or undesired (toxic) effects should already have been established. One can—and some authors specifically require this—use these values as orientation for the dose escalation in man. The careful juxtaposition of pharmacokinetic and pharmacodynamic observations should make it possible to select one or several therapeutic dosages with which the new substance can be studied in phase II using a relatively small number of patients.

## Proof of Effectiveness in Patients: Phase II

It has already been indicated that healthy (young) subjects constitute the ideal population for phase I studies of many medicines. It is true, of course, that very toxic substances, for example cytostatics, whose use is justified only in severely ill patients, will be tested at once in a population that is also the target population for the new medicine. Furthermore, patients suffering from a chronic disease who are taking regular medication will normally be excluded from phase I. In this case as well there can be exceptions. One such occurs when the new substance is intended to be taken simultaneously with other medications that are deemed essential.

In phase II the question of the effectiveness of the new drug is raised rigorously for the first time. In general, such studies are carried out with two or three different dosage levels as well as a placebo control group. In this, the simplest, case we are dealing with two- or three-armed studies. In place of or in addition to the placebo "arm" it is also possible to have various dosages of a comparative drug. In the case of serious illnesses such as AIDS, but also with chronic diseases that can be treated and that should not be left long untreated (with severe hypertension, for example), it is not justifiable to leave patients without effective treatment over the duration of the clinical trial. In such cases the new treatment will be compared with an established therapeutic regime. It is also possible to offer all patients a basic treatment—this is what happened particularly in the HIV studies with protease inhibitors—for example Azidothymidine (AZT)—in order to study the effectiveness of the new treatment against a placebo on top of the basis therapy applicable to all experimental groups. An important question concerns the *definition* of effectiveness. Does a new combination of AIDS medications (for example, two inhibitors of reverse transcriptase and a protease inhibitor), as opposed to a treatment based on *only* inhibitors of reverse transcriptase, lead to a prolongation of life? That, of course, is the fundamental question. However, the answer to such a question will take a number of years and require experiments on thousands of patients. Should a treatment be withheld from these patients until its long-term benefit has been confirmed? Again, there are several answers: With a disease like AIDS for which an urgent need for treatment exists, one makes do at first with studying the behavior of surrogate markers.[11] These are arbitrary indicators that according to current medical thinking are linked to the outcome of a disease. In the case of AIDS, or even with other chronic viral infections, the concentration of viral particles in the blood would be such an ersatz indicator.[12] The therapeutic reduction in the number of viral particles below a critical threshold, for example under the threshold of demonstrability with the polymerase chain reaction (PCR), would be considered a therapeutic success, which, if it can be maintained over a period of months, can make it possible to approve a new medicine, even if there is not yet proof of a life-extending effect. One would proceed similarly with chronic illnesses from whose pathophysiology a surrogate marker can be derived. A medicine that inhibits glycosylation or the interlinking of glycosylated proteins *should* prevent or delay long-term complications of diabetes mellitus. To *prove* such a relationship would take years. On the other hand, a lowering of the glycosylation of hemoglobin to normal values could be considered the theoretical basis for considering the medicine to have achieved a therapeutic success. The convincing ability of a medicine to influence a surrogate marker could be the reason for its temporary approval. Such an approval should always contain the requirement for long-term testing of the clinical usefulness of the medicine.[13]

## Anticipating Therapeutic Reality: Phase III

After a successful conclusion of the phase II studies, the new substance is tested in phase III on a significantly larger and more diverse population of patients. These phase III studies should anticipate the later therapeutic deployment of the new substance as precisely as possible. Thus they reflect the therapeutic reality more reliably than the previous phase II experiments. In this case, too, the new medicine is tested in one or two dosages against a placebo and/or against comparison treatments. Usually, several centers participate jointly in such a program on account of the requisite high number of patients (several hundred to a thousand patients per study). They must all follow the same protocol and obey uniform rules in the way in which results are computed and presented. Such studies are called multicenter trials. Of these, at least one must take place in the country where approval is sought. For approval in several countries it is necessary to obtain data from three to five thousand patients, on average, from all three phases combined. About eighty percent of these data come from phase III studies. It is obvious that the costs of clinical development rise steeply with increasing numbers of patients. A phase III study lasting a year with many hundreds or a thousand patients can easily cost thirty to fifty million dollars. A firm will accept such a high risk in their clinical studies only if it can proceed from encouraging data and if a high return on investment is to be anticipated in the case of a successful outcome. Any other position would make no sense from a business standpoint. That this legitimate entrepreneurial point of view can lead to problems from a societal standpoint is an issue that we shall take up in the last chapter.

Corresponding to their various goals and scopes, the time allotments for phase I, II, and III studies are very different from one another: Depending on the type of medicine under study, phase I can take between eleven and twenty-one months, phase II between fourteen and thirty-five months, and phase III studies from thirty-five to fifty-five months. These numbers, which come from the Tufts Center for the Study of Drug Development, were collected in the period 1970–1982.[14] They may not accurately reflect time periods currently required, since the tendency is unambiguously in the direction of shorter development times. Nevertheless, since that time no dramatic improvements have occurred. The entire development time is still on the order of five to six years; only in exceptional cases has this time frame been considerably reduced. Often, development takes even longer, from seven to nine years.

## Criteria for Quality

As with the preclinical studies, certain criteria for quality assembled under the rubric "good clinical practice" (GCP) are also required for clinical studies. The term GCP denotes a collection of regulations and conventions that should ensure that four basic criteria are met. First, the subjects or patients must know precisely what they are getting involved in when they participate in a clinical study. What are the possible therapeutic advantages? What are the risks and dangers to which they are exposing themselves? The patient must be fully informed as to what is to transpire and must give his written consent: Here one speaks of "informed consent." Second, an "institutional review board," that is to say a sort of ethics committee, must be satisfied that the institution carrying out the study is capable of doing so. Third, the responsibilities of the firm carrying out the study must be clearly established. This includes the presence of standard operational procedures (SOPs), basic rules that establish how a clinical study is to be structured, what will be done with the data collected, and how the results of a study are to be evaluated and presented. Finally, within the framework of these rules the responsibilities of the actual clinical investigators, that is, the physicians taking part in the experiment, are set. This responsibility extends from a conscientious certification of the disposition of all leftover samples of a drug, to documentation of the experiments that were carried out, to a summary report of the results obtained and a comprehensive presentation to the above-mentioned institutional review board. Included are the safeguarding of the patients' interests, the definition and observance of the responsibilities of the pharmaceutical firms involved and of the clinical investigators, and ensuring through the institutional review board that medical and ethical standards are met. This system of regulation was first developed in the United States, but it has since been adopted with minor modifications in all countries that play a significant role in drug development, that is, the USA, Canada, the European Union, and Japan.[15] International conferences have been held to try to bring consistency to the drug development process, and some uniformity has been achieved in both the preclinical and clinical areas. Nonetheless, in Japan especially there remain particular requirements that make it necessary to carry out separate studies for the Japanese market.

## Requirements for Drug Approval and Safety

At the end of the development phase all clinical and preclinical data must be gathered in a petition for approval. With this substantial document, approval for a completely developed product is applied for through the appropriate drug regulatory agency for the country in question. In the United States this document is called a new drug application, or NDA for short. Because of the thoroughness and detailed systematization of the regulations for putting together an NDA, this document has become an internationally used prototype for drug registration, though this does not exclude variations in national requirements, for two reasons. In the first place, it represents the demands of the FDA, which for a long time represented the most comprehensive model for a registration document. Secondly, it became even more desirable from a medical and from an economic point of view to apply for the registration of a new drug simultaneously and internationally. According to the most recent estimates, such a document comprises from fifty thousand to a quarter of a million printed pages. According to the requirements of the FDA, this document must contain the following parts:

- Index
- Summary
- A section on chemistry, providing information on manufacture and all related control processes
- A section on validation, packaging, and labeling, i.e., instructions for use of the medicine
- A section on preclinical pharmacology and toxicology
- A section on pharmacokinetics and bioavailability in human beings
- Microbiology (if relevant)
- A section on clinical data

To this are added additional supporting sections:

- The most up-to-date description of clinical safety
- A statistical report
- A tabular display of all case report forms
- Patient information
- The certification of all patients
- All other information

According to a report for the year 1992, the NDAs for new chemical compounds totaled 214 volumes, of which eighty-seven percent was clinical in content. For line extensions, that is, the approval of new indications, or for the registration of new pharmaceutical forms, the applications comprised an additional sixty-three volumes, with seventy-three percent of the information clinical.[16]

## The Moment of Truth: Approval

The examination of applications by the regulatory agency begins with the certification by the FDA that the document has been received and that it meets the formal requirements of the agency (confirmation of filing). Between 1991 and 1994 an average of twenty-four percent of all applications were returned as formally unsatisfactory or incomplete. When an NDA is received, examiners for the individual sections of the application are appointed. These examiners receive the relevant parts of the application and enter into a dialogue with the firm that has submitted the application on the basis of information they have noted. Often, during the examination phase shortcomings are discovered that must be rectified by the applicant. Sometimes, complaints can be satisfied by explanations or by the provision of additional materials. According to the "user fee act" enacted in 1992, which provides for the payment of fees by the applying firm, the agency is obligated to complete the examination process within 180 days.[17] This time period, however, represents a hypothetical time frame that is adhered to only in sporadic cases, for example, in the registration of new AIDS drugs. Generally, the agency has questions or requests for changes, which it presents in action letters. These letters put the examination process on hold until the questions raised have been satisfactorily answered. In this way the FDA can extricate itself from time pressure. The user-fee regulation has led to a marked improvement in the performance of the FDA. Yet to maintain that it has put a complete stop to bureaucratic arbitrariness and incompetence would be an overly optimistic interpretation of the true state of affairs.

In order to apply their resources to favor innovative products, the agency has instituted a system of classification under which an application is placed either in category P (priority) or S (standard). Priority status is given to substances that compared with treatment currently available promise an important therapeutic advance. For example, these could be substances that treat diseases that have up to now been untreatable. On the other hand, a substance could be given priority status merely for showing promise of greater safety or effectiveness. Likewise, a new medicine could be granted priority if it offers modest, but real, progress, for example a less frequent dosage or the avoidance of side effects that though not dangerous are nonetheless burdensome. Finally, compounds can also receive priority if they offer therapeutic advantages to a patient group that has not been sufficiently taken into account, children, for example. All other medicines come into the standard category and thus do not enjoy a preferential share of resources.[18]

For AIDS and other life-threatening diseases the FDA has instituted a process of "accelerated approval."[19] Here it is a question of providing affected patients as quickly as possible with new medicines that can meet a serious therapeutic need. This process is based on the view that patients suffering from mortal illnesses are more disposed to take greater risks with the safety of a drug than would patients for whom there already exists a safe and effective remedy. The process of accelerated approval rests on two principles: close consultation between the drug regulatory agency (FDA) and the firm making the application, and a merging of phases II and III into a single phase. A minimal prerequisite for an accelerated approval is the completion of two "pivotal" phase II studies that give indication of the new substance's meaningful effectiveness against the condition to be treated. The FDA can also require that further, more extensive studies be carried out whose results will become available only after approval but that in the worst case can lead to a revision of the decision to approve the drug.

Before the agency transmits its approval of a medicine by way of an official "letter of approval," a majority of an advisory committee composed of clinicians and other scientists must have voted to approve the new compound. In general, the agency follows the recommendation of the advisory committee, although it is under no obligation to do so.

Other countries' approval processes are similar to those of the United States and will be sketched here only in outline. The European Union (EU) has developed its own process, which seeks to harmonize the separate national approval regulations. In principle, the EU begins with the assumption that a medicine that has been approved in one EU country can, and in general should, be approved in the other countries of the union as well.[20]

This conviction leads in several steps to a centralized procedure. For products in category A, which includes all products manufactured with recombinant DNA techniques, such a procedure has already been implemented. Also included in this category are products created through the controlled expression of genes in eukaryotic or prokaryotic cells, transformed cells in particular. Finally, all monoclonal antibodies and proteins synthesized by hybridoma cells fall into this group. Other innovative products, including new chemical structures that have not yet been approved in any member country of the EU, are collected in category B. For this group the centralized procedure can be invoked. There is, however, no obligation to do so. The central agency that regulates this process is the European Medicines Evaluation Agency (EMEA), presently located in London, and its committees, the most important of which is the Committee for Proprietary Medicinal Products (CPMP). Each of the fifteen member nations currently sends two representatives to this committee. The responsibility of the CPMP extends to the following realms:

- Evaluation of applications that adhere to the "centralized procedure"
- Efforts to harmonize national regulations
- At the request of the commission or one of the member countries, the adjudication of all open questions that touch on the granting or denial of approval in one of the member countries
- Determination of product safety within the framework of risk–benefit analysis
- Advising the commission on the construction of a European system for drug safety (pharmacovigilance)

## The Never-Ending Obligation: Drug Safety

After approval for a medicine has been granted, the manufacturing firm maintains responsibility for overseeing the safety of the particular drug. In the United States the postapproval procedure is strictly and unambiguously regulated.[21] A safety summary, a collection of all safety-related findings that have been observed in relation to the use of a new medicine, is to be submitted no later than 120 days after the submission of the NDA. Furthermore, the FDA requires that all observed "serious" or "unexpected" side effects be reported within fifteen days of being observed. Moreover, annual reports are to be prepared and submitted within sixty days after the anniversary of the date of original approval. The FDA distinguishes various types of side effects: An "adverse experience" is any undesired effect observed in relation to the use of a drug. It doesn't matter whether the drug itself was responsible for the occurrence. "Unexpected adverse effects" are occurrences that are not mentioned in the instruction leaflet accompanying the drug. Finally, "serious adverse events" are side effects that are deadly, life-threatening, or that cause long-term disabilities or result in the hospitalization of the patient. These are to be brought to the attention of the FDA within fifteen days. Again, these effects must not necessarily be drug-related. They have to be reported in any case.

The European regulations are not uniform and are influenced to a great extent by national legislation. To be sure, the recommendations of the Council for International Organization of Medical Sciences (CIOMS), which strives for an international harmonization of reporting requirements for drug side effects, have meanwhile been accepted and adopted by several European countries. All signs point in the direction of a uniform and internationally approved system for drug oversight in the European Union.

## Drug Legislation to 1962

Without a knowledge of the incidents and catastrophes that have accompanied the modern drug industry, the recent legislation that regulates the development, manufacture, and marketing of drugs would be incomprehensible. In retrospect, it is clear that regulations governing the manufacture and use of drugs were established only after incidents had clearly shown the dangers that can accompany drug treatment. In this connection the notorious thalidomide affair is of particular significance. First of all, the deformities caused by thalidomide have left a lasting impression on many people. Furthermore, that experience makes it clear that the old command to physicians that they "above all, do no harm" must also hold for the preparation of new drugs. It was only after the thalidomide affair that the question of drug safety was addressed in a comprehensive and foresighted way. The laws enacted after the thalidomide affair have proved effective: Today, dangerous side effects can still not be completely excluded. However, they can be caught early and therefore remain restricted to a small group of individuals.

In the United States, drug legislation goes back to the beginning of the twentieth century.[22] However, the law enacted in 1906 (the Food and Drug Act) required only that the components of a pharmaceutical preparation intended to be used as a drug be noted on the package. There were no meaningful restrictions on the formulation of a drug or on any additives used. The legislature was content at that time with instructions relating to the absolute quantity and relative proportions in a drug of particular substances such as alcohol, morphine, opium, chloroform, cannabis, cocaine, heroin, as well as several other substances. It seemed then that this was sufficient—at least in the United States—to satisfy a patient's requirements for drug safety. It was only in 1938 that this attitude changed. The cause was an incident with a preparation of sulfanilamide, an incident that today would certainly be characterized as a "drug catastrophe." In 1938 sulfanilamide was seen as a general medical panacea. As a powder it was applied to wounds, and of course orally administrable forms were available as well. A manufacturer wished to make this medicine available in liquid form and to this end dissolved the substance in diethylene glycol. This solvent has a pleasing color—a pale pink—and its sweetish taste is not at all unpleasant. This preparation was marketed with no clinical or toxicological testing. The solvent turned out to be a deadly poison, from which 107 people died. It is not just from today's point of view that such an incident is to be viewed as a scandal: The public outcry that followed led within the same session of Congress to new legal requirements for drugs. At the center of these laws was the requirement that the drug be safe if used as prescribed by the manufacturer. These instructions for use could contain warnings such as "can lead to addiction." The manufacturers were compelled to provide complete instructions for use for their medicines in an accompanying leaflet and in appropriate places on the containers and packaging in which the drug was marketed. During the Second World War and immediately afterward, drug research was seized by two technological advances: microbiology and biochemistry. Many new substances were discovered, characterized, developed, and eventually marketed. The instructions for use for many of these new substances were too complicated to provide complete and unambiguous information for laypersons in the labeling on the accompanying leaflet or directly on the package. New legislative initiatives took this situation into account. In the Durham–Humphrey Act of 1951 many medicines were freed from the requirement that information about the use of the drug be contained in the leaflet or on the package. These medications could be dispensed and used only by a physician's prescription; that is, their use was placed entirely under the physician's control. The packaging of such drugs and the accompanying leaflet bore an indication to this effect.[23]

Once again, a drug disaster turned out to be the cause for an even more far-reaching law: In 1962 it was learned that thalidomide, a sleeping medication developed by the German firm Grünenthal and tested in the United States by the Cincinnati firm William S. Merrell, had teratogenic properties. The substance had become popular in Europe and especially in Germany on account of its allegedly good effectiveness and excellent lack of side effects, at least in the short and medium term. The discovery of the connection between thalidomide and the appearance of characteristic deformities will be described below. In Europe, deformities due to the clinical use of the substance had already appeared in a large number of newborns whose mothers had taken thalidomide during pregnancy. In the United States, thalidomide was still at that time in clinical trials, and thus the number of birth defects to have appeared was correspondingly small. In retrospect, it has become clear that the refusal of the FDA to approve thalidomide in no way rested on a better understanding than that which the German or other European regulatory agencies could claim. The specialist responsible was simply allowing herself a great deal of time in dealing with the application. However, she took action after the first grave moments of suspicion in Europe were revealed and after American researchers also had encountered sporadic cases of deformities in the upper extremities (phocomelia) in connection with thalidomide. An examination of the situation at *this* moment confirmed the already long-suspected connection and thereby provided the basis for the denial of approval in the United States. In any case, this example shows that delay—for whatever reasons—in the assessment of drugs sometimes can lead to better results than a too rapid course of action.

The full dimensions of the thalidomide catastrophe led again to a public outcry in all the countries involved, particularly in the United States, which, however, was not the most severely affected. Where previously drugs had been judged with growing affection by the public, now the pendulum swung the other way. *All* drugs were now under suspicion. It was also clear that the legal requirements in existence to date were not sufficient to protect the public from new and dangerous side effects of drugs. In the United States, the legislative consequences of this catastrophe was passage of the Kefauver–Harris amendments in 1962. These laws required all drug manufacturers to allow their production facilities to be inspected.

They made manufacturers responsible for their products to a degree hitherto unknown. From now on, a new drug could not be marketed until documented proof of its safety and effectiveness had been provided. Furthermore, the regulations mentioned earlier regarding the manufacture of drugs were introduced—collected under the rubric "good manufacturing practices." Medicines that did not adhere to these regulations were regarded as adulterated and removed from the market. The advertising of prescription drugs was put under control of the FDA, and it was forbidden with increasing strictness to make advertising claims that were not scientifically documented. In addition, the Kefauver–Harris amendments contained many directives still in effect today for the submission of an NDA (we have already touched on this) as well as for granting exceptional approval for the clinical testing of "investigational new drugs." The laws put into effect in 1962 laid down the fundamental regulatory system for the testing and approval of drugs as we know it today. One could say that the way in which drugs are regulated by the FDA and other agencies familiar to us today has its basis in the laws and guidelines that were enacted in reaction to the thalidomide catastrophe.

But development does not stand still. A considerable expansion of drug legislation has occurred since 1962. Above all, the basic framework of the laws enacted in 1962 has been continually refined and modernized, thereby both complicating and to some extent bureaucratizing the development process. But there can be no doubt that the strict regulatory framework that has been erected since 1962 has increased the quality and reliability of drugs and thus has done the serious, research-oriented industry more good than harm. Of the more recently enacted legislation in the United States, the Orphan Drug Act is worthy of particular mention.[24] Unfortunately, the European Union has been unable to pass a similar law. With the passage of the 1962 regulations the process of drug development became very expensive. The pharmaceutical firms no longer felt that they were in a position to finance such expensive developments for medicines against rare diseases. It was precisely this situation that the U.S. Congress wanted to ameliorate with the Orphan Drug Act. The law gives a producer who offers a new or even an already known substance for treatment of a rare disease and who can produce for this particular use the necessary scientific and technical documents in the form of a normal new drug application (NDA) exclusive rights for seven years to the drug in question for use against the disease in question. In addition, certain tax advantages accrue. Rare diseases in the sense of this law are those that affect fewer than

200,000 patients per year in the U.S. It is also possible for less rare diseases to obtain "orphan" status if the manufacturer believes that it has no chance to recoup its high development costs through sales in the United States.

This law has turned out to be a blessing in the United States. Many medicines that originally qualified as orphan drugs have turned out later to be effective for other indications (cyclosporine, interferon-α (Roferon), calcitriol, and many others). They began life as orphan drugs and later found broader application. Thus the orphan drug path has been of help not only to people suffering from rare diseases, but often to a much larger group of patients as well.

## The Thalidomide Affair and Its Consequences

In 1954 thalidomide was synthesized by a pharmacist who had been hired by the firm Grünenthal as a chemist. At that time, Grünenthal was not a traditional pharmaceutical firm like, say, Merck, Schering, or Bayer. It was much more a cosmetics firm that after the war had attained a certain prominence and importance through participating in the antibiotic boom that had begun after World War II. Grünenthal was the first German firm to manufacture penicillin G. The lack of pharmaceutical experience, scientific competence, and medical culture under which the company suffered at the time undoubtedly contributed to the incident that became known in the history of drug research as the thalidomide affair.[25] The molecule that Wilhelm Kunz had synthesized with the idea of finding new ways for synthesizing peptides was classified by Herbert Keller, a pharmacologist working for Grünenthal, as having a structure analogous to that of the barbiturates. This comparison was far-fetched. Even at the time it lacked a scientific basis. However, it gave rise to the idea that the new compound might be used as a hypnotic agent or tranquilizer (sedative). The molecule synthesized by Kunz appeared as a mixture of the optically levorotatory and dextrorotatory forms. In this form thalidomide had no sleep-inducing effect. In the classical tests for evaluating sedative effects, which rest on the correction of posture (the righting reflex or the holding reflex used by Franz Gross and his coworkers at Ciba), the substance had no effect. The workers at Grünenthal attached little importance to these negative results. They were more inclined to doubt the predictive power of the traditional tests than the properties of their new compound, and they designed their own experiments, in which they measured the motor activity of mice. At the time such tests already existed: For example, it was possible to measure the motor activity of mice by placing photoelectric cells and photoelectric beams in their cages and then registering the frequency with which a light beam was interrupted by movement. To describe the elaborate, outlandish, and bizarre experimental setup devised at Grünenthal to solve this problem would lead too far afield. All the same, the team at Grünenthal succeeded with their method in showing that thalidomide did indeed limit the motor activity of mice. Quite arbitrarily, Kunz, Keller, and their supervisor, F. Muekter, assumed a fifty-percent reduction in motor activity as being equivalent to a sleep-inducing effect. The corresponding required dosage was figured by the workers at Grünenthal to be 100 mg per kilogram of body weight. With such a figure, the substance suddenly seemed very attractive to the team of Kunz, Keller, and Muekter. They believed that they had found a compound for which in mice there seemed to be no lethal dosage and whose hypnotic effect was on the order of that of bromides and barbiturates. Later, much of the laboratory documentation with which the Grünenthal team claimed to have established the properties of thalidomide disappeared under suspicious circumstances. A research group at SmithKline and French that later checked the claims of the Grünenthal laboratory were unable to establish any such effect in mice, even with a dose of 5000 mg/kg, which was fifty times the alleged sleep-inducing dose. How could the workers at Grünenthal have gone so wrong with regard to the effectiveness and safety of their molecule? An explanation is surely to be found in their scientific inexperience. Their acceptance of a hypnotic effect was without any reasonable foundation. Moreover, they were working with a racemic form. Later, a group of British researchers were able to show that both the dextrorotatory and levorotatory forms of thalidomide produced definite toxic effects. The best one might say of the Grünenthal team is that they were suffering from wishful thinking: They hoped to have found a substance with unambiguous hypnotic effect and without acute toxicity. To this day no one else has made such a claim for thalidomide. A less friendly interpretation must take into account a certain lack of competence, and most certainly an amazing lack of scientific curiosity. Wilhelm Kunz later admitted that he had noticed the increased effect of the optically pure isomers. Yet no one at Grünenthal had thought to follow up on this phenomenon.

In 1955 the clinical testing of thalidomide began in Germany and Switzerland. In retrospect, all that can be said about these tests is that the alleged sleep-inducing effect of thalidomide on human subjects was never proved unambiguously. Nonetheless, the substance was marketed, in Germany, in 1957, as Contergan; in England, the following year, as Distavan; and in Canada, where it was sold by the American firm William S. Merrell, as Kevadon. Ironically, in its initial clinical use it was considered particularly safe. In some countries it was even sold without a prescription. The impression of relative harmlessness—which in the case of a sedative or hypnotic is not to be taken for granted—grew stronger when in the summer of 1960 it was reported that twenty patients who had taken overdoses of the medicine either in suicide attempts or accidentally not only survived but recovered without complications or lasting effects. Consequently, thalidomide was considered a particularly convenient and safe sleeping pill.

## Thalidomide in the United States of America

In 1960, application for approval of Kevadon was made in the United States, but it was never granted. Frances Kelsey, the specialist assigned to carry out the review of the NDA, dragged out the review process. At the outset, she judged the dossier to be incomplete and at various points ordered the applicant, the firm W.S. Merrell, to provide additional information, particularly more comprehensive data supporting claims of safety. In retrospect, it is difficult to determine whether her complaints were based on genuine errors or whether she judged the NDA itself to be incomplete on the basis of the criteria in effect in 1961. All that is clear is that she harbored no specific suspicions, only that she was unable to make a decision on account of an insufficient quantity of supporting data. The first indications that something was wrong came when Frances Kelsey read a report in the *British Medical Journal* in which the appearance of peripheral inflammation of superficially situated nerves was associated with the use of thalidomide. Such a side effect had not been mentioned in the application to the FDA. In the dialogue between the FDA and the applicant that followed, Merrell offered results on the appearance of polyneuropathies; the FDA demanded more data; the collection of data required more time and delayed the evaluation of the application. This process had not been completed when the teratogenic effects of thalidomide finally became known. Now, of course, there was no longer any question of approval of the application. Frances Kelsey became a national hero and received a high honor for having saved the American public from the thalidomide catastrophe that had now become manifest in other countries. What had happened in the meantime?

## The Riddle Is Solved: Lenz and McBride

Two physicians, Dr. Widukind Lenz, in Germany, and Dr. William McBride, in Australia, came independently to the conclusion in 1961 that thalidomide was responsible for the increasing frequency of birth defects that had been observed since the medicine had been introduced. The deformities involved varieties of a developmental anomaly, known as phocomelia and seldom previously observed, in which the hands or feet grow directly from the shoulders or, respectively, the hips. While in general, these deformities affected only one extremity, they appeared bilaterally in the alleged instances connected with thalidomide. They appeared mostly in the upper extremities and presented various degrees of severity. In some cases the entire arm was missing, while in others it was only the loss of the radius and thumb. Often these problems were connected with deformities of the heart, kidneys, or intestinal tract. The two physicians informed the firms involved, McBride the English license holder, Distillers, in June 1961, and Lenz the German firm Grünenthal in November of that year. The data given to these firms must have awakened the strongest suspicions of a connection between thalidomide and birth defects, even though statistical proof was lacking. However, the possibility that the frequency of deformities in some countries might have other causes could not be excluded. Under these circumstances the firms hesitated to take the necessary steps. Thereupon, Dr. Lenz presented his findings at a conference of pediatricians in Düsseldorf.[26] A week later an article appeared in the newspaper *Welt am Sonntag* in which Dr. Lenz's suspicions were repeated and a demand was made for the withdrawal of the drug from the market. At this point Grünenthal reacted immediately. A few days later, Distillers, the British manufacturer, followed suit, withdrawing the medicine. In the United States, Merrell withdrew its application, while in Japan the medicine continued to be prescribed and sold until September 1962. The use of thalidomide led to birth defects in thousands of children in forty-six countries. In Germany, ten thousand children were affected; there were one thousand in Japan, four hundred in England, just under three hundred in Scandinavia, and two hundred in Canada. There were sixty cases reported in Belgium and twenty-five in the Netherlands. That the United States didn't come away unscathed (ten cases were reported) was due to the fact that Merrell had begun its clinical trials and had not excluded women of childbearing age.

The recall of thalidomide was accomplished under relatively dramatic circumstances. Two days after Grünenthal removed the product from the market, the German health ministry issued a carefully worded statement in which Contergan was named as an important factor in the cause of the increased incidence of phocomelia. The warning addressed therein to women of childbearing age not to take thalidomide appeared on the front pages of newspapers and was publicized by radio and television. In the weeks and months following the withdrawal of thalidomide, the suspicion voiced by Lenz and McBride hardened into a certainty. Lenz and other authors referred to the particular type of deformities caused by thalidomide. According to McBride, and to Lenz as well, women who took thalidomide in the fourth to eighth weeks of pregnancy were at a greater than twenty-percent risk of bearing children with birth defects. In some retrospective studies the figure was placed as high as fifty percent.

After the teratogenic effect of thalidomide had been established, animal models were studied as a way of proving this and similar effects. The mice and rats who had been given high doses of thalidomide had experienced little toxic and no teratogenic effect. Eventually, Somers showed that thalidomide was capable of passing through the placental barrier of rabbits and that in this species it was capable of causing deformities analogous to those in human beings. Later, the teratogenic effects were produced in primates (rhesus and cynomolgus monkeys).[27]

## A Catastrophe and Its Consequences

Thalidomide, a medicine that had at first seemed particularly nontoxic and free from side effects, had within only a few years exposed the most serious weaknesses in the system of drug manufacture in the industrialized world. The quickest and harshest response came from the American public, who reacted to the serious shortcomings in this area that had now become apparent, even though the United States had been spared a genuine catastrophe. The proof of safety and effectiveness of drugs was now universally demanded, and this demand became anchored in legislation. The proof of any teratogenic effects on an appropriately selected species was given particular weight. In 1962 a law had been established that a medicine whose application for approval had not been responded to by the FDA within sixty days received automatic approval. That law was now rescinded. The entire framework of drug development and approval as we know it today was at that time anchored in the Kefauver–Harris bill (Nr. 87-781).

Moreover, an additional circumstance developed, one that cannot be framed in legislation and regulation: The naive trust that before the thalidomide affair had been placed in drugs, and especially in the re-search-based drug industry, had been destroyed, and to this day that trust has not been restored. From now on, mistrust would be the order of the day. Even indisputable advances in drug therapy that have led to thera-peutic benefits, such as the introduction of beta-blockers, calcium antago-nists, benzodiazepines, new immunosuppressives like cyclosporine, and many other medicines, have been accepted with an arm's-length interest at best that has never been free of skepticism. From today's standpoint, the thalidomide catastrophe was the result of a rare mixture of incompetence, lack of accountability, and naiveté. It took place not at the forefront of technological and scientific progress, but in its wake. But in the eyes of the general public, who either do not choose to make such distinctions or are incapable of making them due to a lack of scientific knowledge, progress had shown its dark, seemingly uncontrollable, side. Collective feelings of this sort last a long time. Even the unquestionable progress in drug re-search and therapy that has ensued, even the conscious and strict avoid-ance of further catastrophes of similar scale through strict legislation and through an industry that on the whole acts conscientiously, have not laid these feelings to rest.

## Development as Process

At this point we would like to describe the course of drug development as a complex, interdisciplinary process. We began with some quantitative in-formation. Phase I lasts (with variations that depend on the drug under development) about one year, phase II at least twice that long. Phase III usually requires two to four years. Another two years are consumed by the examination and approval by the FDA or other appropriate drug regula-tory agencies. This means that we can estimate eight years for the total de-velopment process. To be sure, this time period can be reduced, in the ideal case by about half. We shall return to this point later. However, as will be described, achieving such reductions involves assuming a greater level of risk (Figure 5.2).

The total cost for research and development (R&D) of a successful drug was about $116 million in 1976. By 1987 this figure had reached $287 million, and in 1990 the Pharmaceutical Manufacturers Association (PMA) produced an estimate of $359 million.[28] Since that time, figures as high as $500 million have been given (see Chapter 7). Of course, these costs include all costs for discontinued projects and for substances whose

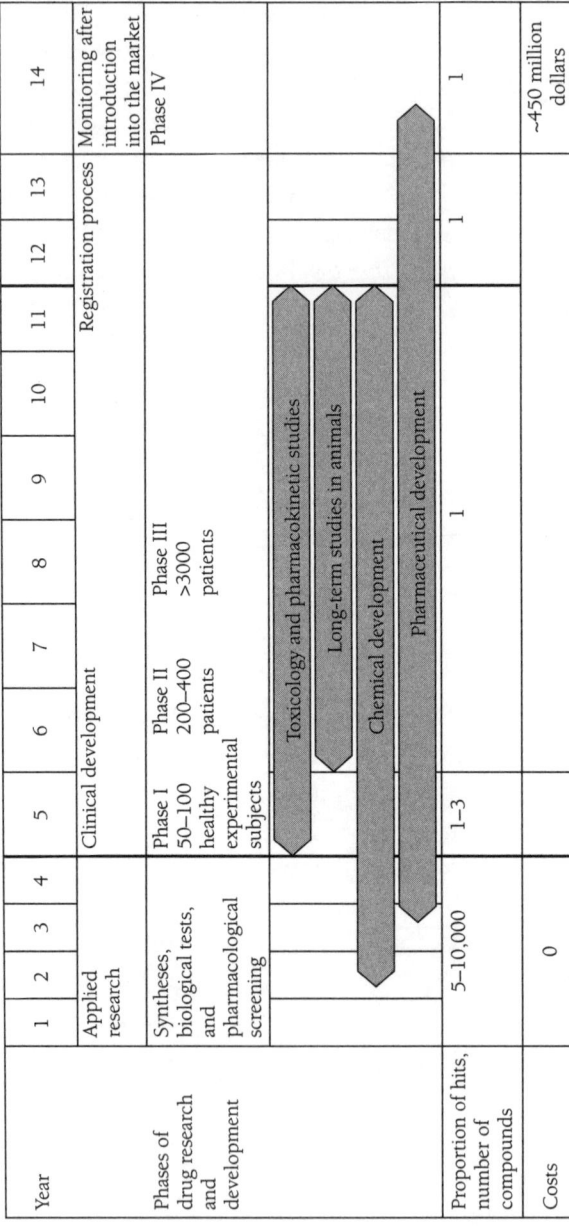

| Year | 1 | 2 | 3 | 4 | 5 | 6 | 7 | 8 | 9 | 10 | 11 | 12 | 13 | 14 |
|---|---|---|---|---|---|---|---|---|---|---|---|---|---|---|
| Phases of drug research and development | Applied research | | | | Clinical development | | | | | | | Registration process | | Monitoring after introduction into the market |
| | Syntheses, biological tests, and pharmacological screening | | | | Phase I 50–100 healthy experimental subjects | Phase II 200–400 patients | | Phase III >3000 patients | | | | | | Phase IV |
| | | | | | | Toxicology and pharmacokinetic studies | | | | | | | | |
| | | | | | | Long-term studies in animals | | | | | | | | |
| | | | | Chemical development | | | | | | | | | | |
| | | | | | | Pharmaceutical development | | | | | | | | |
| Proportion of hits, number of compounds | 5–10,000 | | | | 1–3 | | | 1 | | | | 1 | | 1 |
| Costs | 0 | | | | | | | | | | | | | ~450 million dollars |

Figure 5.2. Development of a new medicine: Time requirements, phases, and costs.

development is not pursued. Furthermore, the figure includes opportunity costs, that is, interest that would have accrued had the capital invested in R&D been invested elsewhere. This cost increase between 1976 and 1990 is nevertheless impressive, especially since inflation has been excluded by giving all figures in terms of 1990 constant dollars. The increase in research and development costs is due primarily to an increase in development costs. This, in turn, reflects the ever increasing complexity of the development process (Figure 5.3). The number of scientific studies that form part of a new drug application (NDA) have doubled: In 1977–1980 the average number was thirty, which grew to sixty in 1989–1990. The average number of patients per study rose in the same time period from fifteen hundred to thirty-five hundred, and the number of pages in an NDA increased from an average of forty thousand (1977–1980) to almost ninety thousand (1989–1992). As we know, the number has grown further, as already mentioned. In many cases it has reached a quarter of a million.[29]

## Speed Up or Slow Down?

Up to now we have been talking about the individual steps of the development process: the preclinical studies, the long-term toxicological studies, and the various types of clinical studies. This deliberately incomplete presentation has focused on the various disciplines, that is, on formulation, biochemistry, clinical pharmacology, toxicology, and the several clinical disciplines. These "disciplinary" studies represent more or less linear extensions or completions of the facts developed during the research phase. Many of these studies run sequentially because the results of an earlier phase are seen as prerequisites for beginning a well-ordered experiment based on the results of that earlier study. Of course, this is true in principle, but only in principle. In practice, many questions can be resolved in parallel. For example, at the beginning of development, that is, about a year before the new compound is to be used in clinical studies, one can prepare enough of the substance to suffice for all the toxicological studies and most of the clinical trials. The pushing forward of the critical development steps or the semiparallel conducting of clinical studies is known as front-loading. In practice, this menas that one waits only for safety data within a time period of several weeks or at most several months before beginning the next clinical study with an enlarged body of questions. If the development process is viewed as an interdisciplinary, but logistically and conceptually unified, process through which a hypothesized

### Distribution of Costs in Drug Development

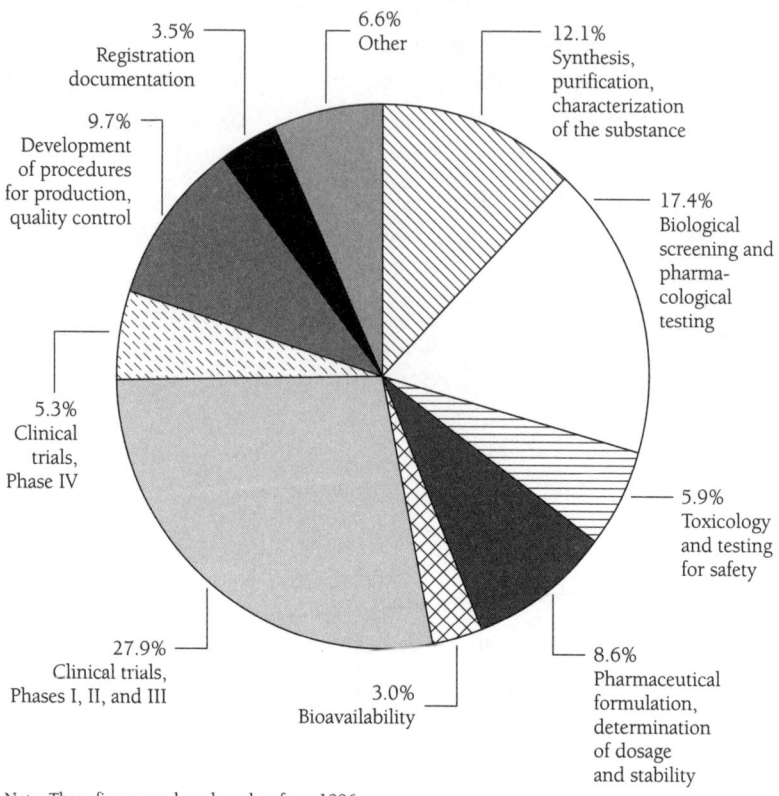

Note: These figures are based on data from 1996.

**Figure 5.3.**   The graph depicts the different categories of costs in the development of a drug. Only a small portion of these costs is attributable to actual research (see the text). (Hardman, J., et al., *The Pharmacological Basis of Therapeutics*, 9E (1996), McGraw-Hill, reproduced with permission of The McGraw-Hill Companies.)

product becomes an actual product, then it is possible to save a great deal of time: The sequence of development steps can change; many steps can simply be omitted, while in some cases two steps, for instance the clinical phases II and III, can be run operationally as a single phase. It is at once apparent that such adjustments that are designed to save time are also risky, and the more original the substance under development, the greater the risks. But a modern development is characterized by precisely this elasticity, the appropriateness of the chosen methodology to the situation at hand. When a promising compound with a new mode of action is developed, the tendency is toward a step-by-step, cautious, procedure. Development is very expensive, and before one puts $100 million on the line, one would like to have as many experimental and, if possible, clinical indicators that the compound will succeed. If one knows or is reasonably confident that with respect to a particular substance one is several years ahead of one's competitors, then a longer development process becomes affordable. However, the situation can change completely in head-to-head competition or if an attempt is made to enter a market in which there is room for only two or three competing products. In a similar vein, a large market will provide a greater incentive to take on certain technical and scientific risks than a small market. A large market offers at least a small chance to recover the invested finances with high returns. In a small market, represented, for instance, by a serious but rare disease, even success will not result in high revenues. We must, however, keep in mind that a drug that is at first effective in a small segment of the patient population might extend to related indications and that a substance that at first promised only modest sales and profits might develop as a consequence of its application to other conditions into an economically successful product. The example of cyclosporine has already been mentioned. It was originally developed for the treatment of rejection episodes in kidney transplantations and gradually was found useful for other transplantation conditions and later for the treatment of autoimmune diseases as well. Furthermore, the possible success of a drug should not be judged on the size of the initial market for the indicated condition. The "transplantation market" was small because before the introduction of cyclosporine there was no method for medical immunosuppression. Like a number of other new substances, cyclosporine created a market for itself. This is a thought that does not always occur spontaneously even to today's marketing managers. And who would have thought when aspirin was discovered and developed that almost a century after its introduction it would be earning money in the prevention of heart disease? The beta-blockers were initially developed for the treatment of angina pectoris. Today they are a fundamental weapon in the treatment of hypertension. Some of them have even proven effective in the prevention of first and second heart attacks.[30]

## The Art of the Reasonable

For every new substance that promises several therapeutic applications, modern drug development must, therefore, determine the optimal speed of development, the correct deployment of means, and the correct sequence of envisaged indications. In this regard, technological and financial risks must be weighed as carefully as the competitive situation. Beyond all technological and economic calculations there is the capacity to envision, on the basis of current epidemiological, therapeutic, and diagnostic information, the development of medical practice several years into the future. Without this desire to innovate, to create therapies and not just reconstruct them, even an organizationally and procedurally perfect development system will remain incomplete. The developer must be a visionary entrepreneur who possesses the capacity and the courage to address certain therapeutic questions broadly, even if they lead into more general aspects of health care. In this sense, development is not only a process of optimization, it also represents the capacity to take part in creating therapeutic schemes and anticipating the future. One may certainly make errors along the way, and there is no guaranteed recipe to avoid them. *One* method of avoiding mistakes consists in the most complete possible mobilizing of relevant knowledge. A part of this knowledge is market knowledge, or finance knowledge. Another equally important, if not more so, factor concerns the medical, that is, the genetic and social, bases of disease, the biochemical mechanisms of their development, their treatment, and the technical possibilities of using such knowledge therapeutically. Within the pharmaceutical industry there is now the danger that drug development will divorce itself from its true foundations and become an instrument of a marketing strategy that proceeds from the status quo, that is, strictly speaking, from the past, from what has already been accomplished and is now destined to lose much of its significance.

Figure 6.1. Industrial biotechnological plant for the production of recombinant proteins by fermentation. (Courtesy of Genentech, Inc.)

# 6
## Management of Innovation: Directing Research and Development

We have noted that research and development are very different functions, representing opposing, or at least very different, cultures. We repeat: Pharmaceutical research must work on therapeutic concepts and realize them in principle, that is, in the form of particular prototypic compounds. To this end, drug research must participate in the progress of basic sciences; it must take up or even generate new opportunities and combine them in novel ways. The *novelty of the combinations* is perhaps the true nature of research, or better, of *applied* research. Rigid schemes are an enemy of the new. The new, however, is the goal of all efforts. Exaggerated formalism and adhering too closely to schematic processes are tantamount to preventing good science. Research requires operative freedom, and only in a generously defined and interpreted framework can techniques and ideas be combined to encourage the creation of novel solutions. Flat organizational structures, informality, a certain "disorder" (which is not to be confused with imprecision), spontaneity, the ever-present impulse to enlarge the boundaries of knowledge and the technological scope for action: These are the cultural characteristics of successful research. One is able, then, to plan how a problem or open question is to be approached. Naturally, one can plan the deployment of means and personnel that should attain a particular goal, which can, for example, consist in taking advantage of a new mechanism of action for treating infections with pathogenic fungi. But what the result of all these efforts will be, how the substance that fulfills the therapeutic requirements will look, what will be the particular characteristics of the substance, cannot be planned. Just as little can one prepare for surprises that bring a particular therapeutic intent into a completely new focus. In short, serendipity cannot be planned for. However, one can create an atmosphere in which unexpected and at first unexplained results are taken seriously. Research is characterized by uncertainty, and it can be planned only to the extent that its inherent uncertainty can be reduced.

Conversely, development begins, as we have seen, with a substance about which a conjecture—in part still hypothetical, in any case, however, testable—about effectiveness against one or more indications can be formulated. And the goal is not only falsifying a hypothesis of effectiveness, but its confirmation under a relatively narrow set of conditions. The step from a well-founded—that is, biochemically, pharmacologically, and through animal experiments—hypothesis to its confirmation in human beings, its applicability to a particular disease or to a pathophysiological mechanism, is often intellectually a very modest one. And this step should *always* be kept as small as possible. That is, we should begin development *only* after the size of this step has been minimized. Where we cannot do this, it is better to continue research, sometimes by including clinical or clinical–pharmacological methods. Thus the *conceptual* step from a compound under development to a finished medicine is not significant. Yet the *demand for its reproducibility*, and with it for the reliability of all the methods employed for testing it, is enormous and is often underestimated by those from outside, even from scientists who are professionally not familiar with this requirement of drug development. And it is beyond doubt that the methodological complexity of this step, as well as the stringency that economics require, will never be mastered without planning and without strict formality, without the painstaking observance of particular rules applicable to the process and without the employment of merciless measures of quality control.

## Two Cultures

Does a cultural discrepancy between research and development justify an organizational separation?

### What Research and Development Have in Common

That would probably be going too far. For all their intellectual and methodological differences, research and development remain highly interdependent. It is essential to the success of research that the original substances that it produces have the chance of being developed and whenever possible reaching the market. That is, so to speak, the "material" side of the interests of research. Just as important is the functional side. For a new substance that comes from the research process, the evaluation of the fundamental pharmacological or chemical hypotheses through clinical development is truly the acid test. If these hypotheses are confirmed, then that is also an indirect confirmation or validation of the experimental methods involved. These can be biochemical, cell-biological, or—and this category is especially important—animal-experimental models in which a disease or partial aspect of a disease is emulated. "Reality" is always equivalent to clinical findings, and only they can provide research with a final "reality check," namely, the proof that its experimental concepts and methods stand up to therapeutic reality. But a negative result also has meaning. The fact that a hypothesis that has been supported by theoretical considerations and animal experiments *cannot* be confirmed by broadly applied clinical experiments leads always to this question: Why not? This question, posed at various experimental levels, often leads to answers that make possible new and successful approaches to research. Thus research needs the results of development for its own methodological orientation. Conversely, development is dependent on the logical and consistent use of all information arising from research about a new substance or a new effective principle. Clinical failures can often be explained only with reference to preclinical development. Equally, experimental findings can give important information with respect to the indications that might be clinically explored with good prospects for success. Unfortunately, there is a tendency to grant commercial considerations too much weight in the choice of the direction in which a new medicine is to be developed. For example, a "humanized" monoclonal antibody against the small subunit of the IL-2 receptor proved to be an inhibitor of the T-cell immune response in many experimental procedures. On the basis of experimental findings, an application of this antibody against rejection episodes after organ transplantations or as an immunosuppressive in T-cell-related autoimmune diseases seemed to be within reach. Nevertheless, for purely commercial reasons the antibody was tested first for an indication that although fitting into the conceptual framework, nonetheless possessed neither a fully conclusive theoretical basis nor the relevant experimental findings, namely, graft-versus-host disease. This disorder arises through the reaction of transplantation-borne lymphocytes against the recipient organism. There exists no good animal model for this disease, and results from animal experimentation were not available. The clinical double-blind studies, conducted very carefully, proved a great disappointment.

On the other hand, an experiment carried out with the antibody afterwards for allogeneic kidney transplantations showed—as research had predicted—an excellent clinical effect. The frequency of rejection reactions was greatly decreased from what had been achieved with conventional immunosuppressives.

We shall not enumerate further examples at this point, all of which could show that commercial considerations, or better, *wishes*, that are not supported experimentally have proved to be poor guides for the development of new substances. Results obtained from research have proved far more reliable. Therefore, in choosing the first area of application of a new substance that also embodies a new active principle, those indications should *always* be chosen for which there exists, on the basis of scientific evidence, a high probability of success. Once the substance has been anchored in one area, then it will be possible to open up further, often more lucrative, areas of application. Conversely, when one is saddled with a negative clinical result based on commercial considerations, it is then often difficult to find the support for further experiments, even when they are scientifically well founded.

## Divergent Cultures, Common Goals

Therefore, as we hope this digression has illustrated, research and development are functionally dependent on each other. Furthermore, from a broader, entrepreneurial perspective they are to be considered functions of similar type: Both serve the *provision* of new and innovative compounds. For these reasons we argue here that these two functions should *not* be organizationally separated. Of course, it is possible to have clearly separated research and development organizations. However, they should both— and this is decisive for their success—be housed under the same roof. They should be placed under a single director of R&D, thereby being integrated at least at the managerial level.

The culturally determined centrifugal forces and the centripetal forces based on mutual dependency often balance out. Whether two decisively important organizations will develop toward each other or will grow apart depends often on whether or not they are jointly directed. The experience in several large pharmaceutical houses has shown that an organizational separation of the two realms can also lead to a mental separation or strengthen even more the separation that is always to some extent present. When development (clinical research, drug safety, project management, and regulatory activities) are under divisional management (and with it usually marketing as well) and research under an international research directorate, then management needs to take particular pains to prevent a drifting apart of the two organizations. Experience, and especially the most recent experience in several organizations, shows that the presence and effectiveness of powerful integrative forces exist more in wishful thinking on the part of the firm's management than in reality. Viewed historically, the separation that has been carried out in many places between research and development is the first step by which the classical pharmaceutical industry will rid itself of research altogether and become a purely development and marketing organization. On the other hand, the continual success and functional consistency of a firm like Merck Sharp & Dohme shows that the organizational unification of R&D represents a success factor that should not be underestimated.

If one defines, following Joseph Schumpeter, innovation as a new product or new process at the moment of introduction into the market, then it is immediately apparent that research and development are the decisive steps in innovation. For first the new must be found and precisely described; afterwards, the conditions of its practical application have to be worked out. This must happen as quickly as possible, for all that is new is susceptible to the passage of time.

But if research and development are the decisive steps on the road to innovation, then the question presents itself as to the manner in which these functions should be implemented so that innovations will in fact arise. This approach leads us into the management of research and development.

## Research and Development in the Enterprise

We encounter research in the enterprise in two prototypical roles: One is a fundamental activity that can generate a new company and remain at its core; the other is a "product-generating" means to an end. In general, younger industries frequently develop around new technologies and are characterized by them. We also observe that research in mature enterprises is often used to achieve entrepreneurial goals that have been derived from market requirements.[1] As we have already noted, the pharmaceutical industry developed from two sources: from the nineteenth-century apothecaries that wished to make drugs available in larger quantities and in progressively more uniform quality and purity, and from the dye divisions of chemical firms. In the latter case the chemical enterprises understood the relationship, described by Paul Ehrlich, between dyes and chemotherapeutic agents, and they recognized the opportunity that this relationship offered. In both cases new technologies were in play. However, the application of already available technologies was also important in establishing new entrepreneurial activities. In both cases chemistry played an essential role.

Technological ideas coming out of basic research have influenced the pharmaceutical industry over and over again. We have already mentioned chemistry, in both its synthetic and analytic functions. Through the discovery and eventual development of penicillin (1929 and 1938–1942) many firms were induced to learn the methods of fermentation, purification of microbial agents, and the search for new antibiotics. These activities proved extremely successful and made possible through industrial and scientific contributions significant advances in the treatment of bacterial and other microbial infections. The rise of biochemistry, whose beginnings lie early in the twentieth century, opened new pathways in pharmaceutical research for understanding organ functions, which in the 1950s and 1960s led to a great variety of new medicines. Finally, molecular biology since the end of the 1970s has affected biomedical research so fundamentally that completely new paths could be followed in the search for drugs. At first, the goal was recombinant proteins and monoclonal antibodies and their derivatives. Today, molecular-biological methods and concepts play a leading role in all treatment-oriented branches of biological and pharmacological research. Genetic research arising from molecular biology, that is, the mapping and sequencing of genes, as well as an understanding of their function, their interdependence, and their role in physiology, pathophysiology, and in developmental biology, open to medicine and pharmaceutical chemistry for the first time the possibility of identifying and understanding pathogenic genes and gene products. This approach in turn offers new possibilities for the discovery and development of treatments that act on the *cause* of a disease. Gene therapy, which is still in its infancy but is being pursued with great dedication, could eventually, if it were to be successful, shake the current scientific and entrepreneurial foundations of the pharmaceutical industry and even put parts of it out of business.[2]

## Research Creates New Markets: Not the Other Way Round

The history of the pharmaceutical industry is surely not a history of systematic attempts to satisfy therapeutic needs. Rather, it is a chronicle of scientific breakthroughs that had an effect on the generation of new medicines for specific diseases. Progress in the pharmaceutical industry followed developments in the relevant research areas—not the other way round. In this sense one can indeed speak of a history of medicines, but it would be better to say that medicines themselves have made history, namely, the history of the pharmaceutical industry.[3]

Scientific and technological developments have opened new entrepreneurial possibilities, but in between there were always long periods in which progress was primarily in the optimization of treatment by means of small changes and adjustments. For such modifications the desires of marketing were able to provide the sole impetus. In such a configuration research is truly only a means to an end, that is, to reaching specific product profiles within technical possibilities that were set long ago.

At present we are observing that molecular-biological research in its many varieties is generating new therapeutic concepts for the realization of which many new firms are being founded. Here research is making an appearance as a driving force that leads to the establishment of new firms. The directors of such firms are often scientists themselves; they have an excellent understanding of the technological bases of their entrepreneurial activities. Their main interest is not in setting up marketing plans but rather the rapid transformation of technological impulses into medically relevant products. Today, this new industry, called the biotechnology industry in the United States, is quantitatively and qualitatively already a principal source of important therapeutic concepts and products. Considering the number and originality of the projects being worked on, which use not only biological but also chemical methods and those of information technology, this young industry is in a position to leave the classical pharmaceutical industry far behind. What they are lacking in large measure are the methods and capital for product development. Herein lies a chance for the pharmaceutical industry to redefine its own role as an innovator through strategic alliances with smaller, technologically motivated firms.

## Management of Research

Whoever seeks innovation needs research and development and must also have an idea of how these two functions are to be implemented. The question of management of research and development can be posed on three levels: (1) the organizational and institutional level, (2) the individual level, and (3) the level of process. On the broad, institutional level, we must ask how research and development divisions are to be structured, how their relationship to the organization is to be defined, and how they are to be managed. Of particular interest is the question of how independent the research operation should be, or, put the other way round, how strongly research and development should be integrated into a single enterprise.

In the second category, the questions are different. Here the questions have to do with how goals are set together with scientists, how the course of a project is followed and evaluated, how individual performance is to be compensated, and what "climatic" requirements must be met in order that scientists feel at home and remain productive.

On the third level, the question is whether drug research can be understood as a unified, or at least adequately structured, process, and if so, how this process can be optimized through technological means.

## Research as Institution: The Appropriate "Constitution"

Let us begin with the first category, that is, with the question of the "constitution" of research and development and the "form of government" most appropriate to the nature of these functions. I wish to state once again that drug research and development are very different occupations, which for their optimal advancement have different mental and cultural requirements. Research, basic or applied, is individualistic, even when it cannot function without division of labor. A researcher must constantly call into question existing knowledge and established categories. Thereby, he or she also questions those who establish such categories: themselves, their colleagues, their supervisors, and the very principles of organization without which no enterprise can exist. For example, the division of research into therapeutic areas, in functional or project-oriented hierarchies, represents such organization. Research in the best sense is the process that discovers new insights from a chaos of possibilities, findings, and concepts, insights that have consequences in the real world and thereby prove their worth; and therein lies an important difference between drug research in a pharmaceutical firm and basic research at a university or a research institute. In these latter cases the demands for proving the value of a result in the world of medicine, with its rigid criteria of effectiveness, safety, and prolongation of life, are less strict. At least they are not as closely scrutinized as in industry.

An activity that creates new realities does not thrive in hierarchy, in formalism, in compulsory regulations, all of which have their origins in quite other areas of experience, for example in the military (Switzerland), government bureaucracy (Germany, France), or in a purely market-oriented occupation (sales organizations, production divisions, and such-like). In fact, research does best when it can organize and orient itself only loosely along functional lines. The best scientists with the best ideas organize themselves into collegial groups that work across disciplines to reach a goal, which can consist as much in purely scientific notions as in ideas that point toward a therapeutic application. Several such groups should be directed by an experienced colleague who also comes from research, one who is valued and followed on account of his good judgment, his overview, his vision, and his personal integrity. This is, of course, an ideal. The great fluidity of research ideas and the important role played by lateral thinking and imagination demand ever new collaborative groupings: Research organizations must be, and should be allowed to be, flexible. Hierarchy, bureaucracy, unnecessary diversion on account of the requirements of other areas of the enterprise—these are all counterproductive influences on a research organization. Unfortunately, these influences are on the rise. For a long time the particular attention of leaders of pharmaceutical firms or divisions was directed at the most far-reaching possible integration of functions in all areas of the enterprise, whereby an optimization of transfunctional processes was to be attained.

Sometimes such strivings took on the most absurd forms. According to the formulation of a "reengineering specialist," "everyone should decide about everything." Cooperative work became the big thing. Knowledge of one's field, particularly medical and pharmaceutical knowledge, was writ small. Now, the simplification and acceleration of an interdisciplinary process across functional boundaries is a very sensible idea. However, in carrying out a plan that must meet the demands of the entire organization, the boundaries of logic are often breached. If the universality of the boundaries between divisions and functions is improved at the expense of competence, then in the end what is created is "universal" incompetence. Enterprises, and particularly large enterprises, have a difficult time instituting changes. Many employees must be won over and brought around to a new point of view. Such a necessity often induces a coarse or simplified formulation of the situation. Whether we are talking about "total quality management," "management by exception," "management by objectives," or even just the improvement of processes, there is always the danger that such "new" developments will create their own jargon, which tends to be reminiscent of the manipulative doublespeak of totalitarian regimes. And thus the more sensitive and often the more intelligent employees are driven away. It should be clear that research does not flourish in such a climate of thought and speech. It is characteristic that biomedical research has fared very much worse under totalitarian regimes than it has in open societies.

For a while, such notions had some success, but managers forgot that researchers, and precisely the most talented and productive among them, are often individualists who find such constraints anathema. In practice, this means that good scientists will abandon firms that no longer offer them freedom; they will prefer other professional possibilities to work in a pharmaceutical firm.

## The Temptation of the Small Enterprises

In the meanwhile, a biotechnology industry has sprung up—primarily in the United States, with more than fourteen hundred small enterprises—that offers, particularly to the clever and successful scientist—alternatives to employment in the pharmaceutical industry, particularly so, since the founding or management of a biotechnology firm can be combined in many cases with an academic career. An additional important factor comes into the picture: The pharmaceutical industry throughout the world is at present in a phase of great change and consolidation. These events are connected with a chronic disparity between the size of the enterprise and their expenditures on the one hand and productivity on the other. Later, we shall go into this in greater detail. These events were often accompanied by layoffs, even in the domain of research and development. What for decades had been a particularly attractive feature of the pharmaceutical industry, job security, particularly in research and development, has suddenly been cast into doubt. If such stability, combined with good work opportunities and a widespread liberalness in the choice and treatment of research topics, no longer exists, why should researchers look at all to industry for employment? There is more bureaucracy, more hierarchy, less autonomy, and less long-range or future-oriented thinking than are to be found in universities or in the biotechnology industry. On the positive side, however, there existed until recently a great deal of stability and a generous number of positions. If these positive factors disappear or are greatly diminished, then many pharmaceutical firms will find themselves in difficulty, because the most talented and most highly motivated part of the new generation could well choose alternative career paths. One cannot avoid noticing that it has become increasingly difficult over the last few years to encourage talented young researchers to enter the pharmaceutical industry. Every research manager can tell such a tale—especially in regard to the situation in the United States. But if the best researchers become unavailable, this would result in a long-term decline in the quality of research in the pharmaceutical industry, with an attendant loss of productivity. Today, the number of new introductions of pharmaceutical compounds has sunk from a worldwide annual total of about sixty (mid 1980s) to about forty in 1994. On the assumption that this development will continue, many firms, perhaps even the entire "established" pharmaceutical industry, would lose their justification for existence as research enterprises and would turn into purely development and distribution firms.[4] Research would then be left to the smaller firms, who in their entirety constitute today's biotechnology industry. They would be supported by the universities, which at least in the United States are reformulating their societal roles and see themselves not only as producers and disseminators of new knowledge, but also as partners in the societal implementation of their inventions and discoveries.

It is beginning to be understood in some pharmaceutical firms that it would be a dangerous development which would institutionally separate research and development from each other. Despite their differences, these two functions need and depend on each other. Ideally, they should continue to dwell alongside each other under the same roof and be under a single director. It would also be difficult for pure development organizations to acquire the most promising projects: Biotechnology firms will always prefer to work with pharmaceutical companies that on their part can offer something in research competence.

## A New Structure for Research

If in the future a firm wishes to maintain research and development as the pillars of the enterprise, then it must follow new paths in establishing a research environment within its own house. The best enterprises of the biotechnology industry, and academic departments and institutes as well, can provide important orientation in this regard. Let us not forget that researchers require a certain amount of independence and flexibility in the selection of goals. The nature of their work and their thinking requires informality. The minimum in hierarchy commensurate with the size of the organization should be established by those who themselves come from a scientific background and who possess the trust of their coworkers. This trust should extend to the conviction that the integrity of the scientific method and the methodological and medical integrity of research projects will be protected by the research director. Intellectual stimulation from other scientists is another important element to be provided. Moreover, researchers should be protected from having their energies diverted into nonscientific and nonmedical channels. Finally, scientists who join a firm, be it a biotechnology firm or a pharmaceutical company, are interested in obtaining a commensurate share of success, including financial success. Bonuses can fulfill this requirement. But the most effective methods are financial arrangements that signal a shared ownership, a stake in the firm's success. In this regard stock options are attractive, since researchers can affect their value over the long term. Not only do they offer the possibility of financial success far beyond the nominal salary, but within different limits for individual scientists, they also represent the opportunity to become involved in determining the fundamental direction of an enterprise. And even if this influence is minimal, to work or participate in effect in one's "own" firm carries more weight than work in an alien enterprise to which a scientist offers his or her services strictly in exchange for money.

## The "Semiautonomous" Center

The changes to be sought come to this, that research units of large enterprises be made independent, that they be given the status of independent laboratories or institutes, which themselves bear the responsibility for success within certain agreed-upon boundaries. What powers must the larger enterprise retain for itself? A few: First, the firm must decide together with its research centers on long-term plans and strategic directions. Second, the provision of an annual budget, which should be bound to the formulation of goals for the fiscal year, should serve as an instrument of direction. Third, the research management of the firm must see to it that the work of the various centers (if there are several) should proceed in such a way that duplication of effort is avoided and that opportunities for synergy are taken advantage of. Finally, large joint projects and sharing of resources that affect the strategic orientation of the firm should be dependent on approval by the directors of the division or enterprise. It is not only desirable, but in the given circumstances absolutely essential, that a system of financial incentives and bonuses is created in which the ownership of stock or stock options plays an important role. Thus equipped, a research center of the pharmaceutical industry would be thoroughly able to compete with the biotechnology industry and university institutes.[5]

Within rather wide boundaries, the scientists would be responsible for their own success, and scientific success would be tightly coupled to financial success. The manner in which this success is to be achieved—whether by going it alone or by cooperative work—would be the determination of the research center itself, as also the manner in which joint work is to be carried out and the everyday work rules. A success precisely defined as to quantity and quality would be the only determining measure by which the enterprise judges its research centers. What should this success look like? In the first place, there would be a particular number of substances (of a loosely defined degree of originality) that are offered to the firm per time period (year) for its own development. The research and development leadership of the firm (pharmaceutical management) would decide whether and to what extent use would be made of these offers. Bonuses would be paid only for substances that are so attractive that the firm actually develops them. It is a matter of debate whether for such projects further "premiums" should be paid to the employees when the projects reach or pass critical points in the development process and eventually are marketed. An important question touches on the "degree of ripeness" of such development projects: Should they be simply pharmacologically and toxicologically, chemically and galenically, well-characterized products, for which clinical proofs of effectiveness and safety are still outstanding, or should such a clinical "proof of concept" always be included?

There is much to be said for working on a project until proof of clinical safety and efficacy has been obtained. However, this is not always possible. In the case of rheumatoid arthritis or a medicine against multiple sclerosis, a proof of effectiveness can be produced only within the framework of a regular development process. Short clinical studies can merely permit predictions about the safety of a medicine, about possible dosages, about its kinetics, or about its influence on particular surrogate markers. Where such results are relevant to the decision as to whether to put a substance into development, they should be obtained along with the experimental data. Alternatively, if the threshold against development can be crossed on the basis of the experimental data alone, clinical studies would not be undertaken. Nonetheless, the research centers should have the methodological prerequisites for carrying out clinical studies of Phase I and the beginning of Phase II. In those cases where the sponsor firm elects not to develop the compound, the center should have the option to offer the substance under license to other developers.

The detachment of research from the confines of the company as outlined here should make possible a climate that is more conducive to research, and it should also contribute to the effect that scientists who are responsible for the future of an enterprise will not be sidetracked by long meetings and by being drawn into unproductive activities. However, this should not hinder or interrupt important communication between research and development, production, and marketing. It is not a question here of moving research into full autonomy, but of bringing it carefully into a balanced "semiautonomous" situation. At the moment, at least, it appears that the only alternative to such a step lies in the gradual surrender of research by the pharmaceutical industry, which would then limit itself to development and distribution.

## People and Goals

The management of researchers as individuals should be guided by two parameters: by the goals of the enterprise and the personalities of the employees to be managed. The responsible manager—and this holds at all levels of management—must know the goals of the enterprise and the concomitant strategic principles and be able to identify with them. Of course, he must have a precise idea to what extent, in what manner, and in what time frame his department, or indeed the entire center, must contribute to reaching these goals. The second point touches on the personality of the employee being managed, his capabilities, his strengths and weaknesses, and his personal ideas about himself and the firm in which he works. Within the framework of the goals of the firm—and eventually outside this framework as well—management means the attempt to help the employee to achieve his best and most creative possibilities. The intellectual processes that lead to the discovery of new facts and to the invention of new methods cannot in the final analysis be planned. They can only be recognized by their results. What we call management, that is, setting an example and providing an orientation, is here applicable only indirectly—transmitting to some extent general intellectual and entrepreneurial values. Intellectual demands, honesty, tolerance, informality, and constructive criticism are some of the values that a research director must transmit to create a climate in which unusual ideas and original solutions to problems can thrive. These are principles that can be of use in other places in the enterprise. However, they are almost always required in combination with other criteria such as loyalty to the enterprise, orientation toward customers, or a focus on quality. To such criteria there is in princi-ple nothing to object. But they operate on a plane that has nothing to do with intellectual creation of something new.

## Setting Strategic Goals

In research it is a matter of creating—within particular thematic bound-aries—something new, not yet thought of, as yet unproved: a matter of *cre-ating*, that is, not of producing, applying, or selling. One can hardly com-mission a scientist to come up with a brilliant idea, or even a clever one; but one can bring together individuals who stimulate one another intellec-tually. One can create a climate in which the unusual has a chance to be discovered, an atmosphere in which all members of a group are bound to-gether by a delight in new insights and solutions. The critical realm in which an enterprise's reason for being meets its goals and partial goals on the one hand and the personalities of the individual employees on the other consists in the formulation of the research themes to be worked on. This occurs in most cases in reference to an already existing historical back-ground. Thus the choice of large areas such as cardiovascular research, in-fections, or research on the central nervous system will be oriented around existing strengths, that is, around historical conditions, around medical needs, and around the total strategic alignment of the enterprise. Does a firm wish to establish itself in the area of general indications—against all competition—or does it think about particular indications, rare diseases in which with some research successes one might play a dominant role? Fur-thermore—and this is often forgotten—the choice of broad areas must also be made according to whether they are scientifically and methodologically sufficiently developed. There are many examples of vast sums being spent on inaccessible areas with no tangible successes having been achieved. The "crusade" against cancer, thrust into prominence with huge financial outlay by a naive and poorly advised American president, was a washout: The sci-entific basis for the invention of new treatments was at the time (beginning of the 1970s) unavailable. The completely unsuccessful research by many firms in the area of metabolic disorders illustrates this point just as well. Until very recently there were only vague ideas about the causes of insulin resistance or the genetic bases of obesity or of diabetes type II. Only now have these areas become accessible—through the identification of genes that contribute to diseases, on the one hand, and through the elucidation of receptor-dependent signal transduction on the other.[6]

## Research Goals in Detail

Once the major target areas in the dialogue between research and the firm's management have been set, then the formulation of the operational details, that is, the creation of original and potentially successful research projects, is for the most part the concern of the scientists themselves. Certainly, on the basis of a drug that has been successfully marketed, concrete wishes for a follow-up medicine can arise by means of which the one success can be extended and perhaps built upon further. In such cases, which can certainly make considerable strategic sense, the marketing department can play an important role as a stimulus in the search for new drugs. However, true pharmacological innovations have up to now always been the result of speculative research projects that set out directly from knowledge gained from basic biomedical research in a somewhat unpredictable way. Furthermore, the pursuit of chemical structural relationships has played an important role in the creation of new drugs. In this connection we recall the connection between sulfanilamide and the diuretics or also the sulfonyl ureas.[7] Such activities take place, of course, far from the realm of market considerations. An enterprise that makes the error of setting its research goals through market-oriented wishful thinking instead of letting them be directed by scientific goal-setting within specific thematic boundaries is playing with its very existence. For practically all important breakthroughs in drug research are based on the perceptive recognition and exploitation of chances created by basic research. Aspirin, antibiotics, muscle relaxants, beta-blockers, calcium antagonists, all the way to the interferons, prove the truth of these statements with stark unambiguousness. Marketing can formulate therapeutic needs only in terms of "today's" vision. If one considers that the research that makes such a vision of today possible already lies many years in the past, then it becomes clear that the wishful perspective of the marketing department often relates to products and formulations that may be useful today. In five years time, however, they can be obsolete. The broad setting of detailed research goals through scientifically defined mechanisms should, of course, not be used as camouflage behind which research and development separate themselves from the requirements of the enterprise. It is the concern of research management to see to it that in the research projects selected there reside not only the principles of good and current science, but also the characteristics of successful drug research. The goals established must, based on present knowledge, be attainable within a reasonable period of time, that is, within several years. The costly search for low-molecular-weight antagonists or even for agonists of a protein receptor whose natural ligand is likewise a large protein runs counter to this expectation, except under specific circumstances.[8] Just as important for drug projects is that they offer reliable starting points for chemical considerations, at least to the extent that the moment at which chemistry can be brought into play lies not more than one to two years in the future. From the point of view of research management, the scientific process of setting goals must also include a certain balance with regard to the estimated time periods and the risks to be taken into account, as well as the necessity of fitting the number of projects that will eventually be worked on to the available resources. Often, industrial research groups spend many years with exploratory projects, that is, with research projects that are perhaps interesting but that are also very speculative and very risky. Such activities are important because they can open up new territory. In general, universities and research institutes offer better conditions for the development of such ideas than do industrial laboratories. Therefore, it would be advisable to develop exploratory projects primarily or even exclusively in cooperation with such entities. It goes without saying that for this the right partner must be found. We shall discuss below how this situation might look in detail and what effects it can have on the productivity of the firm's own research.

The greater the creativity of the research organization, the clearer it is that a considerable portion of successful research management consists in choosing from among various projects and project ideas. In this, several criteria are always to be considered: the scientific quality of the project, its prospects for realization, the estimated time to complete it, the personal and financial expenditure, and the risk of failure. In the larger framework of the enterprise as a whole there are questions to be answered about the possible therapeutic and commercial uses that can result from the success of a project. These general criteria, especially in the evaluation of very original projects, should be used with great restraint. Their application would make sense only if the results of research projects could be predicted with precision. However, as we know, this is seldom the case.[9]

## The Tools of Research Management

The regular evaluation of research projects, the establishment of new goals, and ensuring that goals are reached should be carried out according to the principles of peer review. For an institute or research center this means that all group directors and department directors regularly meet under the chairmanship of the research director to analyze and evaluate the projects currently under study. For the cohesion of the group it is necessary that projects and suggestions that arise be presented by the scientists themselves who are responsible for them. The most expedient arrangement is that a regular cycle be followed, in which every project is fundamentally evaluated at least once, preferably twice, each year. This does not, of course, exclude ad hoc evaluations that arise when special problems come up. Such a committee within a research center should decide on the allocation of resources for projects and on their continuation or termination, even, on occasion, on the acceptance of new projects. They should see to it that the decisions made are understood by all employees and supported by them as much as possible. However, we are not talking about management by consensus. When unresolvable differences of opinion arise, the responsible research director has the duty to decide and to justify his decision to higher management.

When an enterprise runs several research centers, it can be an advantage to have an international committee to which all directors of research centers or institutes under the management of the director of global research belong, and this committee will be responsible for the formulation of strategic directions and for the regular critical evaluation of all areas of work. The preclinical development functions and the leader of international clinical research should also be involved in such evaluations. The participation of the manager responsible for regulatory matters as well as that of the director of project management is desirable at least for questions relating to development. In firms that have globalized research and development, each research center will be responsible for its own particular areas of research; that is, it will have its own thematic focus. Since the acceptance of a new substance from one of the centers into the research portfolio is one of the most important decisions that regularly occur in a pharmaceutical firm, the international committee should decide based on the recommendation of the center responsible for the project in question. If clinical research, project management, and regulatory matters are joined together in a self-contained development operation, then the director of this operation has an important voice in deciding on the acceptance of a new product into development. In some cases the divisional management reserves the last word in deciding all such cases.

## Peer Review: How Science Governs Itself

For the evaluation of projects at the local level, external experts should also be asked for their evaluation. While the company scientists must make the final decision, they should benefit from the advice of their academic peers. Many companies hesitate to allow such external scientists a complete view into their projects. The advantages of such evaluations— which are in distance and objectivity and in a comprehensive evaluation of a therapeutic area, taking into account the competitive situation—are more significant than the dangers that might exist in the revelation of confidential information. Competent scientists who also serve as experts for other scientific bodies, for example the Deutsche Forschungsgemeinschaft or the NIH, know full well the limits that are to be observed, not only in the interests of the firm for which they are consulting, but also in regard to the integrity of the peer review process itself.[10]

The evaluation and control mechanisms here described should be complemented by thorough and comprehensive reviews of selected areas of work and indications every one to two years. These reviews should be of an interdisciplinary character, and thus they should include, in addition to preclinical scientists and clinicians, members of the marketing department. Through such a broad evaluation and agreement, a perspective across the entire company can be achieved that simplifies communication between R&D and other divisions of the company. The conclusions of such synoptic evaluations will then be discussed among the research management, in some situations with company management as well, and where indicated will be translated into action.

Of course, through the activities of such committees as we have described, new projects or even new areas of activity may arise. Actually, the creation of new ideas and projects is not the task of committees, but that of each individual researcher. The research organization of a large enterprise should be a marketplace of ideas to which each member of the company can contribute. When ideas in this environment receive sufficient support, then they will obtain access to one of the existing decision-making committees and thereby have the chance of achieving the status of a project. As much as the creation of new scientific possibilities is the task of each individual researcher in the company, the research directorate must see it as their particular duty always to alert company management to new strategic possibilities—that is, to the chance at entering into new areas of activity and new technologies. Research management consists not only in "directing research" and the evaluative and selective accompaniment of work related to projects; its task also consists, and perhaps primarily, in recognizing future technological and scientific developments and in the design of company-specific future strategies.

## Research as Process

New technologies have the potential of fundamentally altering the process of pharmaceutical research. In spite of the division of labor involved, this process has up to now remained individualistic. A project began with a chemical or pharmacological idea that fit into already known or hypothesized biochemical and/or pathophysiological contexts. Such an idea could be very specific, concerning, for example, the blockade of a receptor or the inhibition of an enzyme: The work of Sir James Black on beta-adrenergic receptors and histamine type-2 receptors provides examples of the blockade of a receptor. The identification of carboanhydrase inhibitors based on sulfanilamide illustrates the second category. There have also been more global approaches to research: Cyclosporine was discovered in the search for microbial metabolites capable of suppressing the formation of lymphoblasts in a mixed lymphocyte culture. There was almost always a mechanism in the foreground that could be brought into association with a disease or with important physiological or pathophysiological processes. Blind screening was also carried out, to be sure, but it contributed little to the great drug breakthroughs at the close of the twentieth century.[11] Of course, this assessment depends on how one defines blind screening. Here we shall define it as the attempt to test as large a number of substances as possible, without any sort of mechanistic concept, in several, opportunistically selected biological tests and to take note of accidental "hits" as potential substances to be developed.

Successful projects therefore grouped themselves mostly around concepts, hypotheses, and speculative ideas. At the beginning either a chemist or (more frequently) a biologist or a team from both of these key disciplines stood at the center. A project obtained structure, in the sense of processes strictly organized on the principle of division of labor, only as it showed increasing possibility of success. Now this could all change. Genome research will soon make available the structures of all the genes of higher organisms. In many cases we are able to use databases containing genetic information with all relevant annotations as we use libraries.

We know, for example, that the insulin receptor is phosphorylated after binding of its ligand and that in this state it sends a biochemical signal. Its dephosphorylation by a specific phosphatase terminates this signal. An inhibition of this phosphatase should prolong or strengthen the insulin signal. In this way insulin resistance which is typically found in diabetes mellitus type II, but also in obesity and in certain forms of hypertension, could be overcome. The identification and characterization of such enzymes could be extraordinarily simplified by the use of "gene libraries," and in fact, drug research has already profited from the mapping and identification of genes. Many structures (more than thirty percent) of the genes that have been sequenced have still provided little indication as to their function. And even where a rough functional classification is possible, this does not often result in a convincing indication of a *specific* function. Once we know the full gene sequence, we can also express it. This means not only that the DNA sequences are available to us, but also the corresponding proteins. In this connection, in analogy to the genome, one has also spoken of the *proteome* to denote the entirety of proteins produced by an organism. On the basis of particular structural indicators it is possible to classify many of these proteins as membrane receptors, cytoplasmatic receptors, ion channels, tyrosine kinases, metalloproteinases, phosphatases, cytoskeletal proteins, and as members of many other functional or structural groups of proteins. And it is possible in large measure to test the capabilities of these proteins to bind small molecules.

New techniques of combinatorial chemistry make it possible to build rapidly very large substance libraries, containing hundreds of thousands or even millions of individual compounds.[12] In the past these libraries consisted primarily of peptides or nucleotides of various sequences, but more and more frequently they are built of the same building blocks that an organic chemist traditionally uses for drug synthesis. But many of the libraries that have been created suffer from too little diversity: The large number of substances are too similar to one another. However, we shall more frequently be seeing collections of substances with a high degree of diversity, which can be tested against a likewise very large number of target proteins in a highly automated testing process employing robots. To be sure, what will be tested first is "binding." Whether this binding leads to functional changes must be tested first in simple systems and then step by step in ever more complex ones. But there are logical and self-contained ideas about how to proceed from simply binding substances to compounds that bring about a gain or loss in function, a change in intracellular localization, or a dimerization of proteins. The next step should consist in the selection of those substances and target molecules whose interaction can be formulated in a functional and/or pathophysiologically relevant context, and these molecules then would be substances to "optimize." This could occur by way of combinatorial techniques or by syntheses oriented toward structure–effect relationships.[13]

## "Automated" Drug Research?

Most of these processes should be able to be automated. With the compounds obtained in this manner one could then proceed to animal experiments in order to select suitable candidates for development. Perhaps the sequence of events sketched here sounds a bit hypothetical and somewhat oversimplified. But there can be no doubt that computer-directed, highly automated processes will play an increasing role in drug research, just as they are doing in other areas of biomedical research. This means that the individualistic character of pharmacological research will be displaced across a wide swath by methods whose hope of success lies solely in the mastery of large numbers of tests. The unifying conceptual work will be carried out by a few highly valued employees. It is obvious that such strongly process-oriented research requires additional tasks for management that are highly reminiscent of corresponding tasks in the realm of drug development. It is also conceivable that the extensive reorganization of research into a process would require a different type of employee. If chemists and biologists whose interest was in specific structures or in the mechanism of action of drugs have so far dominated drug research we could now see the need for a new type of scientist. In the new situation a vertical knowledge or the deep understanding of a particular problem is not so important, but rather the complete use of all available information for the discovery of as many candidate substances as possible. Perhaps engineers, computer scientists, and physicists will be more in demand than medical chemists or pharmacologists. However, at the end of all these "processes" there will always have to be scientists who can join the individual discoveries into a total picture; thus both will be necessary: the "engineer" who thinks in terms of processes and the scientist who studies particular structures, signal pathways, or mechanisms.

It is not to be expected that robotic and automated processes will come to dominate research immediately; there is also the possibility that such work will be contracted out to small firms that specialize in it. And yet, the influence of a way of thinking that is less oriented toward biological and pathophysiological mechanisms and more on automated processes will make itself felt more strongly in research; there would be an associated increase in planning activities oriented toward meeting quantitative and qualitative norms, deadlines, time-critical stages, decision points, and the like. The task of a research manager, which until now was oriented on the one hand toward the goals of the enterprise and on the other toward the establishment of conditions that served the nurturing of individual creativity, would be supplemented by a third, process-oriented, element.

## Management of Innovation

Until quite recently the task of a research director in the pharmaceutical industry has consisted in employing the research process to identify the greatest possible number of original substances in-house and to make a large number of them available to be developed by the firm itself. In principle, this is still the case. In a certain way this task has been made easier by the creation of interdisciplinary research units in which biologists and chemists work together to create new treatment possibilities in a therapeutically or operationally defined area. Earlier, chemists and biologists worked in separate departments; today they are for the most part members of the same department. In the future the research director with responsibility for a particular area such as oncology or cardiovascular diseases will not have the task of preparing compounds arising exclusively from within his own firm. Such a perspective no longer suffices in a tightly networked scientific world, in which new information that can be relevant to one's own particular area is continually made available in a great variety of areas. An individual responsible for research in an area such as oncology or heart disease must attempt to obtain for his firm the best ideas, the best research methods, and also, finally, the best substances for development. In this milieu, *where* these new research ideas or substances for development come from is not so important. Thus the responsibilities of a modern research manager extend beyond looking after his own research. He must know and be able to evaluate the worldwide state of drug research in his area. Furthermore, he must be capable of evaluating the relevant basic research, at least in broad outline. Finally, it will be his task to establish from his own research, in cooperative work, and eventually also through financing research outside his own firm a program that over the long run will ensure his company a competitive position in his area of responsibility. He is therefore not only a research manager, but in a very real sense a manager of innovation for a particular area, for example for oncology, cardiovascular medicine, or infectious diseases. To be able to fulfill this role he must also have access to the necessary tools. If one wishes to react quickly and flexibly to opportunities for cooperative work that offer themselves from without, then one cannot remain within an inflexible budgetary framework in the old style.

## Cooperation Demands Flexibility

Theoretically, a budget is set for in-house research as well as for joint work with third parties or for the financing of external work. In practice, however, it is impossible to decrease sharply the costs in internal research as soon as joint work with an academic partner or a biotechnology firm is started. The division of resources for work in one's own company and for joint work with a third party can vary within wide limits. On average, among the leading pharmaceutical firms fifteen to twenty-five percent of the research budget is spent on cooperations—or as one prefers to say today, "strategic alliances." This tendency is definitely on the rise.[14] To create the necessary flexibility, research managers should have the possibility to transfer part of their research budget into an account from which they can at any time withdraw funds for financing joint ventures. Unfortunately, international accounting rules prohibit unspent funds from being carried over to the next fiscal year. However, this obstacle can be circumvented. This is an important requirement for greater flexibility at the level of research managers who are responsible for specific areas. A research manager who directs, say, the oncological area should agree on goals for quality and quantity with his head of research and the divisional management. With what mixture of his own research work and collaborative projects or projects conducted under contract he then reaches these goals should to a great extent be left up to the manager himself. Therefore, he should be able to utilize funds that had been set aside for joint ventures in his own research if this appeared advisable to him.

## From the Research Department to the Business Unit

It is a relatively small step from this more entrepreneurially based view of the function of a research manager to that of the director of a business unit. A business unit is a part of a company which generates its own income. Its success is measured in terms of profit remaining after all expenditures have been subtracted from income. In the case of a research unit, one need not go to such an extreme. Nevertheless, a unit, for example the above-mentioned oncology department, could be organized to function almost like a separate firm. It would then have to have available its own clinical pharmacologists and clinical oncologists, as well as business professionals (MBAs) knowledgeable in the market for medicines used in oncology. Likewise, it would need its own controller. This last-named function could be shared, as could personnel services (human resources), with other divisions of the company. The unit would find new medicines and develop them up to the point of "proof of concept." The term "proof of concept" denotes critical clinical experiments in Phases I and II which help to confirm the fundamental pharmacological hypothesis. This expression can also be used in a more limited way. Sometimes it merely describes the clinical confirmation of a single effect or the absence of critical side effects. If the development of a new antibiotic would depend on a long half-life, one would accept this parameter in healthy subjects as proof of concept. Likewise, the absence of a particular side effect typical for a class of substances could be a *conditio sine qua non* for the development of a new substance. Thus the proof of concept would consist in the demonstration that this side effect really did not exist. Of course, such noncentral, but important, components of the effect are interesting only if all the remaining positive effects that one expects from the substance are present.

The company—or better, its pharmaceutical division—would then draw up a contract with the respective division setting out the conditions under which substances are taken into global development after proof of concept. It should also make clear how a business unit deals with substances that cannot or should not be developed by the firm itself.

How far one goes along the road to independence of research units is always a question of the firm's culture, that is, the historical contingencies and the personnel involved. It is certainly more invigorating to transfer more independence and entrepreneurial responsibility to smaller units. On the other hand, it is possible to go too far in this direction. The entrepreneurial "autonomous" structures must still create a unified whole among themselves and with the company at large. When business units that actually represent research areas cease to work together, either because they see one another as competitors or because they have set false priorities, then at that point they have certainly gone too far. It is also counterproductive to grant entrepreneurial autonomy to units organized on too small a scale, say to project groups: If too many groups believe in their individual missions and place their own success ahead of that of the company as a whole, this can lead only to an atomization of the structure of the enterprise and to serious functional deficits. To find the proper balance between the autonomy of research units on the one hand and their union in the enterprise at large on the other surely belongs among the most difficult tasks of a research director, and to other top managers as well.[15] Many talk in glowing terms of entrepreneurship and the expression of creativity, only in the end to act like petty bureaucrats.

## Development Is Conservation of Innovation

Innovation is a product and a process at the moment of its introduction into the marketplace.[16] Pharmaceutical innovations, according to such a definition, would be new medicines that have already been approved and that have entered several markets. The process by which a pharmaceutical innovation comes into existence can be divided into many stages: Basic research; creation of a concept; translation of an idea of a mechanism into biochemical or cell-biological tests and animal models; the search for chemical lead substances; improvement and optimization of a lead substance according to criteria of effect, bioavailability, stability, and metabolism; identification of the substance to be developed; clinical testing for side effects; establishment of a presumably effective dosage; tests of effectiveness on patients with the disease to be treated; broad clinical testing; and creation of a final document that can serve as the basis for approval. Fundamentally, however, there are really only two steps: the creation of a new substance and its development to a medically usable medicine. The first step is a matter of research, the second the concern of development.[17]

## Novelty Is Time-Sensitive

A new substance that has shown its effectiveness in a variety of experimental models and for which data are available regarding its bioavailability, pharmacokinetics, and toxicity that indicate a possible clinical use—such a substance, about which it is further known that it can be produced in uniform quality and purity and that it can be formulated into a medicine, represents something new, at least it should be clearly recognizable as offering the possibility of a novel or significantly better treatment for an important disease. However, such novel substances are rarely found just lying around. Usually, the new compound represents just one possible solution to a problem with which others are also occupied. That is, usually there is competition, and the ultimate success or lack thereof depends not only on the originality and attractive properties of the new compound, but also on whether it is brought to market quickly. Novelty is time-sensitive. What is new today can be run-of-the-mill by tomorrow or the day after. Thus it is not sufficient to create something new. It is also necessary to preserve the novelty of what has been created. Development is the preservation of novelty against the leveling influence of time. It is, after creation of a new substance, the second essential component of every innovation. Therefore, it must be in the interest of every pharmaceutical firm to set up the development process to be as rapid and efficient as possible. In the case of a successful preparation, saving a year in development time can represent the equivalent of several hundred million dollars in earnings. On the other hand, during the development process one must be on the lookout for unexpected phenomena that can indicate unknown but exploitable effects of a substance. The antidepressive effects of the tuberculostat iproniazid were first discovered in the clinic. The systematic investigation of these unexpected effects by interested clinicians, who were at first supported only halfheartedly by the sponsoring firm, led eventually to the important class of monoaminooxydase inhibitors, which are used to this day in the treatment of depression. As another example: The antipsychotic effects of the phenothiazines were discovered not by pharmacologists, but by clinicians. Without creative clinical observations the possibility of finding and developing new diuretics would never have become obvious from the use of prontosil. From sulfanilamide, the antibacterial metabolite that is derived from prontosil, there leads a pathway to acetolamide, to the hydrochlorothiazides, and finally to furosemide, which never would have been found without the stimulus of clinical observation. This list could be continued. There are at least a dozen similar examples.[18] Equally important is the systematic survey of particular clinical side effects, which taken together can perhaps give indications of unexpected toxic properties of a new substance that did not appear in the animal experiments. Therefore, not only must development be quick and efficient, but it must also be carried out attentively, with intelligence and creativity. It is difficult to accommodate simultaneously these two competing desiderata.

In today's environment, speed and efficiency of development have the clear upper hand over scientific thoroughness and the discovery of new therapeutic possibilities. This is also reflected in the composition of development organizations. Well-educated physicians are moved into the background in favor of project managers who, while they oversee the complex process of development, often know little about the medical and scientific aspects of the substances being developed under their supervision. Also, the influence of financial analyses, which usually lead to the computation of a "net present value" for a new preparation under development, has grown much stronger than it was in the 1950s and 1960s, when many medicines were discovered by observation of clinical side effects. These analyses can be useful for orientation purposes. However, very frequently they rest on superficial epidemiological considerations and are almost always incorrect. Therefore, they should not be the decisive criteria for terminating further development of a new drug. Unfortunately, in some companies we are experiencing excesses that do not bode well for their future. In the effort to speed up the development process, many companies follow the principle known as front-loading. The term describes the earliest possible initiation or completion of tasks that can otherwise adversely affect the speed for the entire process, for example, the production of sufficient quantities of the substance for preclinical toxicity studies and clinical tests in Phase I. With front-loading, the functional interdependence of specific processes is sacrificed, for example, the relation between the amount of substance required and the results of preclinical toxicity or metabolic studies. One produces enough of a given substance so that everything that can be carried out in parallel from an operational perspective is *indeed* carried out in parallel. Instead of proceeding sequentially, one works in parallel wherever possible. Time is thereby saved, though of course at the price of an increased risk of higher total costs. It is plausible that front-loading makes sense if the clinical profile of a new substance can already be fairly accurately predicted (indication, dosage scheme, expected effect, avoidance of side effects). It is just as plausible that this strategy is risky and that it is most suitable for cases in which a preparation in an already known category of substances is being developed, perhaps with the goal of a precisely defined therapeutic improvement. This could be, for example, an antifungal substance with the same fungicidal properties as amphotericin B but without nephrotoxicity, a calcium antagonist that produces no reflex tachycardia and no ankle edema, or a cephalosporin with a broad spectrum of effectiveness and a particularly long half-life. However, the concept of front-loading becomes problematic in all cases in which truly original substances come into development, that is, prototypical substances whose clinical application is in no way well-defined. There are additional examples that could be named that are valid at least at the time of writing (1997). How, for example, should an endothelin receptor antagonist that blocks ETA as well as ETB receptors be employed? What specifically could an endothelin antagonist add to the already existing therapeutic repertoire? Coronary spasms (Prinzmetal's angina) would be one possibility; vasospasm associated with cerebral hemorrhage or after cerebral trauma would be another. Chronic heart failure? Here the same objections hold as those already mentioned for the treatment of hypertension. In such cases it can be much more productive to determine the best indication by carrying out small, informal studies before cranking up the activities that cost serious money. This is particularly applicable when several candidate substances are being investigated and when the most suitable of them is to be chosen on the basis of clinical results. Front-loading would not be a recommended strategy in such a case. Such considerations also apply to new cytokines. It is well known that interleukin-12 produces a cellular immune response. Likewise, the formation of interferon-γ as a result of stimulation by interleukin-12 and the role of interleukin-4, whose formation is inhibited by interleukin-12, are all well-known facts. These facts, however, leave many possibilities open for clinical application of interleukin-12: treatment of tumors, viral infections, allergies. In such cases is it better to carry out multiple studies for several hypothetically justified indications or to gamble everything on one throw of the dice, if possible on the one that seems the most profitable commercially? In this situation, the majority of scientists would decide on a search strategy followed by a rapid development of the compound for the "best" indications.

Businesspeople often think otherwise. For them a new medicine is "good" only if it is directed toward an economically profitable indication. They do not readily get involved with a search strategy with uncertain outcome that even if it is not very expensive, nonetheless consumes a great deal of time. "Either it works for chronic heart failure better that all previous drugs, or else we'd better steer clear of it," is a typical statement about a substance under development. Of course, one can try to do too much at one time, and many managers fear this with good reason. But in a short-sighted and all too narrow-minded striving for profits one can miss out on good possibilities whose exploitation calls for at least the first two of Paul Ehrlich's five requirements for success: intelligence, patience, health, luck, and money. The secret of success does not require that the researcher always have his way, although this would be better than if businesspeople would assert their authority over too broad a front. The nonscientists who are responsible for the company must oppose a certain critical resistance to new ideas, but it must be an elastic resistance, an intelligent resistance, that forces the researchers to examine their arguments with care and to support their respective cases with their best ideas, and above all with data. Resistance to new ideas cannot ossify into a cost-schematization that in the end has more to do with ideology than with the duty of an entrepreneur to look after his enterprise.

## Do Research Costs Threaten Innovation?

Research and development are very expensive. Moreover, in pharmaceutical firms, research costs are growing more rapidly than all other costs: The inflation factor for research estimated by the United States government has been at around fifty percent higher than the overall rate of inflation.[19] On top of a general increase in expenses of four percent there must be added an additional two percent for research costs. Why is this so? The main reason has to do with the continual improvement in research methods. Methodological, and frequently technological, improvements compel research units to invest in newer, more efficient, but also more expensive equipment within relatively short time periods, that is, every few years. Research is becoming more complex, more efficient, and more costly. If today you have an NMR apparatus running at 400 MHz, then tomorrow you will need—just to keep up with the competition—a bigger, more powerful piece of equipment running at perhaps 600 or 700 MHz. The development of transgenic animals (transgenic mice, "knock-out" or "knock-in" mice, etc.) and thus of animal models on whose germline quite specific genetic alterations are made requires not only time, expensive equipment, and sterile rooms for keeping animals, but above all, as all experimenters in this area know, it requires space. If you require access to hundreds or indeed thousands of transgenic animals (usually mice), you will have to freeze embryos. To have available a particular breed of mice or even several such strains, you will need to implant embryos and breed the animals in large numbers. This technique has proved itself as one of the most elegant methods in biology and drug research for studying genetic function.[20] The necessary expenditures are, however, considerable; in many institutes the availability of units capable of providing transgenic animals has already become a limiting factor for progress in research.

The list of such examples could be prolonged at will. The introduction of robots for carrying out repetitive tasks, new miniaturized techniques of analysis, microchips on which cDNAs are fixed and with which the genetic expression of particular populations of cells (organs, blood cells, etc.) can be measured under a variety of conditions and as a function of time, new techniques for mapping and sequencing genes: All these innovations have taken hold of drug research during recent years, and they are continually changing it. Research is thereby becoming more efficient, but also more expensive, and those who can't afford to participate in this technological progress will be forced to limit themselves somehow, for example, working in fewer areas or posing narrower questions employing specific technologies. Of course, this has its limits, especially when development costs are likewise rising.

The cost of finding and developing a new drug has risen from about $24.4 million (1956–1966) to $54 million (1976), $231 million (1987), and $359 million (1990). Today, the costs are in the range of $350–500 million for a new chemical entity (Figure 6.2).[21] Of course, these costs are not simply the direct costs for a particular project. They include the costs of unsuccessful projects as well. Furthermore, the direct costs make up only about sixty percent of the total costs given here. About forty percent is allotted to indirect costs, that is, to lost income for the capital invested in research. Of course, there are particularly expensive developments and relatively inexpensive ones. The development of a medicine that slows the course of a chronic inflammation like rheumatoid arthritis, for example, is more difficult and protracted than the development of an antibiotic. Nevertheless, the discovery and development of a new medicine that will be introduced worldwide still costs several hundred million dollars. A firm that spends $600–700 million each year for research and development and whose research costs rise every year relative to inflation can expect with such expenditures to bring at most two new drugs to market. An investment for research and development of $700 million today (1997) will, with a cost increase of six percent per year, be equivalent to research and development costs of 1.25 billion dollars in the year 2006! In a steady state that works with a six percent cost increase, the introduction of two new substances per year requires a doubling of the annual research and development expenditure of $700 million over ten years.[22] Since research and development costs must be financed from pretax earnings, and what is more, cost increases for medicines are subject to strict national controls and will not be permitted to reach six percent per year, the pharmaceutical industry finds itself in a dilemma. It must attempt to improve its production of new medicines: That is, it must do more at lower cost. Within certain limits this is possible. Every process can be improved and made more efficient and economical. The pharmaceutical industry concentrated at first on development, which in any case has accounted for an ever-

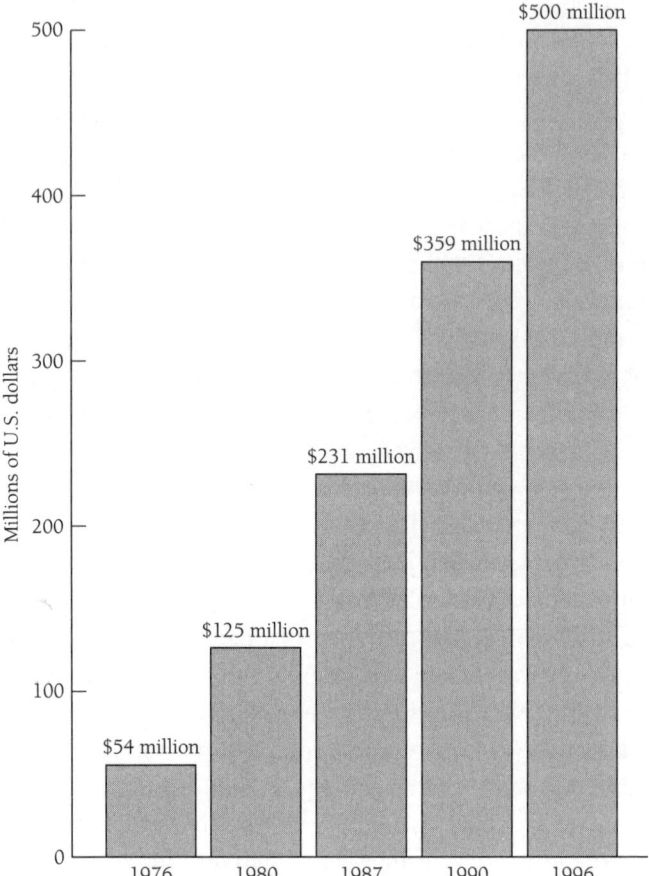

**Research and Development Costs for a New Medicine**

Figure 6.2.    Costs for the discovery and development of new drugs between 1976 and 1996. The numbers come from a variety of analyses, which were carried out with different methods and therefore are not directly comparable. The numbers have not been adjusted for inflation. All numbers also contain the costs for failures in research and in development. They also all contain both direct costs and investment costs. In each case, the interest rate in individual calculations was taken to be eight or nine percent.

increasing portion of the total research and development costs since the middle of the 1970s. In the 1950s and 1960s the development costs made up on average somewhat less than half of the total research and development expenditures. They have risen almost everywhere, until in some companies the development-to-research cost ratio has reached 80:20. The greatest contributor to this increase in development costs has been the disproportionate rise in expenditures for clinical trials. The cost-cutting in health services in various countries has led to the result in recent years that these costs are not rising with the same dynamics as previously. Moreover, firms have learned to tighten their development process, to concentrate on the important questions for a registration of a drug, and to save time. Development times on the order of four to five years (from introducing the substance into clinical development until registration in the first country) are quite common today, and in many cases (Rocephin®, Risperidon®) the time period has been reduced to three-and-one-half years. These are impressive numbers when one thinks that ten years ago the development times were more often on the order of seven to ten years.

The research and development costs have risen sharply both as a percentage of sales and in absolute numbers over the past twenty years. If research and development during the 1970s and early 1980s represented eleven percent of sales on average for the internationally active industry, today that figure is sixteen to seventeen percent. For a few firms the research and development costs come to more than twenty percent of sales. The uncertainty that is presently ruling the health-care sector has led to a conservative stance in the research-intensive pharmaceutical firms in planning their spending on research and development. Some firms predict that their expenditures for research and development will rise considerably in the coming years—from a current thirteen percent to over twenty percent of sales. No one can say today where the optimal relationship between income and research expenditure lies. But the increasing complexity of science and technology and the fact that more detailed information is continually required in order to produce a truly innovative product point to the likely result of increased research costs. Society's demands for effectiveness and safety of medicines point in the same direction. There is no longer a place for "just average." On the other hand, cost-cutting in the health-care sector and growing competition among pharmaceutical firms are forcing the most extreme discipline in controlling costs. Furthermore, research and development costs must be seen in relation to other expenditures. Expenses for marketing and sales traditionally make up about twenty-five percent of sales, often even more. More than ever before, the actual therapeutic value of a new medicine is important today. A drug must be original, effective, and safe. Is it still justified in an intellectual climate that is shaped by these requirements to spend more money on sales representatives, advertising, and publicity than on research and development? Should not the traditional advertising for drugs be gradually replaced by medically based explanatory information accompanying new drugs? What would be the effect of such a change in marketing methods on the costs for sales and marketing?

One might hypothesize that the creation of something new should be permitted to require more resources than the distribution of the newly created. One day, perhaps, an enlightened pharmaceutical industry committed to the goal of providing information that is well presented and scientifically correct will find a more productive relationship between the available resources for research and development and those for marketing than the balance established by today's pharmaceutical industry. Only time will tell.

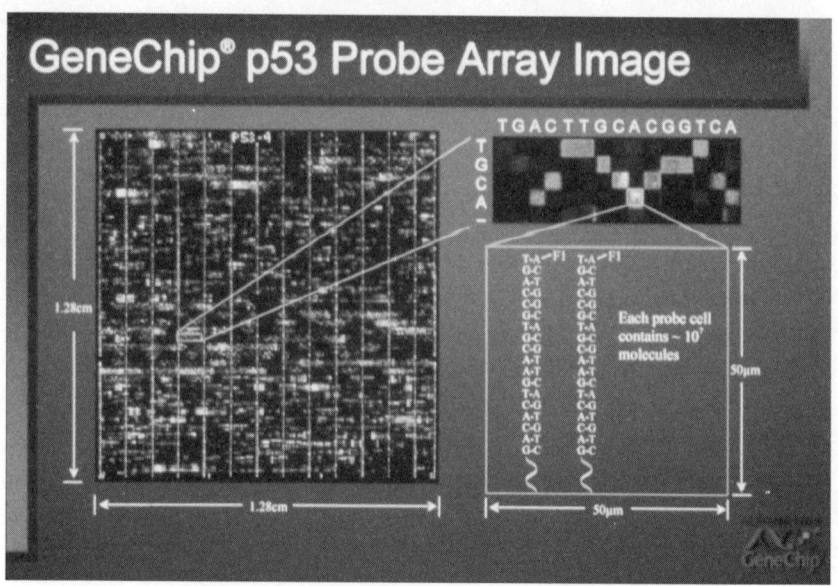

**Figure 7.1.** Gene chips are used for the detection of specific nucleotide sequences within genes. If a fluorescent probe matches the sequence on the chip, it will hybridize to this sequence and send out a fluorescent signal. "Mismatches" will result in weaker binding or no binding at all and the absence of fluorescence. (GeneChip® p53 Probe Array Image courtesy of Affymetrix, Inc., Santa Clara.)

# 7

# The Future of Research and Development in the Pharmaceutical Industry

The future of pharmaceutical research and development depends on many factors, above all on the history of the industry itself. For many decades the pharmaceutical industry possessed a virtual monopoly on the discovery and development of new drugs. Nowhere else were to be found united under one roof all the knowledge and skills, and all the disciplines, that are employed in drug research; and so long as the total costs of health care, and particularly the cost of drugs, seemed reasonable and financeable, there was hardly any reason to take objection to this "monopoly." Now all this has changed. Drug prices, in connection with overall increasing healthcare costs, are viewed critically and are either directly or indirectly controlled. The industry's freedom of movement has thereby disappeared, a freedom that it previously made use of to portray itself and its frequently redundant products in a manner not always quite accurate, but that it also used to search for cures for rare diseases and other health problems in the Third and Fourth Worlds. For now that, too, is gone. The reasons for the current economic difficulties of the pharmaceutical industry are quickly enumerated. They can be arranged in three categories: demographic factors, economic development, and scientific and technological advances. We shall treat these categories in turn. In the industrialized countries there is no longer an excess of births over deaths, the number of births barely managing to replace the number of deaths. Even so, there has appeared an imbalance in the population structure: People today are living longer than they did a decade or two ago. This is leading, at least temporarily, to an increase in population, and in particular an increase in a portion of the population that requires extensive medical care. Moreover, in many countries the demands on the financing of social needs have risen disproportionately to government income. Everywhere, it is difficult to cut back on social services that have been built up over the years. Of course, in Continental Europe the situation is particularly difficult, because the scope of social services has developed to a degree that in the United States seemed unthinkable. Public health services compete with other social services, while at the same time they find themselves under ever greater pressure on account of the growing demands of an aging population. On top of this, economic growth in Europe has, with the exception of Britain, been modest or indeed has stagnated in recent years. In Germany, for instance, the gross national product has grown an average of 1.2 percent per year since 1992.[1] This situation has further narrowed the freedom of movement of governments. Subsidies for obsolete industries (coal mining and the steel industry) maintained for sociopolitical reasons as well as indifference and opposition toward new, job-creating technologies have altered the industrial landscape. Thus the development of biotechnological processes and the construction of a biotechnology industry have been set back at least a decade by the opposition of the green parties and social democrats in Central Europe. Moreover, this has been made even more difficult by unnecessary legislation, which, to make things worse, was often implemented in a very bureaucratic way. High-tech industry has not become sufficiently important in Europe; the number of obsolescent and obsolete industries is high; many foreign investors no longer believe that industry as a whole will grow energetically. It is only in the most recent past that a somewhat more favorable picture has emerged.

A further important point has to do with the great complexity of research and development themselves. In particular, the new biology has significantly enlarged the methodological repertoire of pharmacological research. But also in chemistry significant breakthroughs have occurred: Consider the evolution of combinatorial chemistry along with advances in understanding and utilizing chemical structures. Finally, all areas of research, and especially development, have been influenced by information technology. Drug development, for example, is seen as a process through which various pieces of information obtained sequentially or in parallel are united into a document that will make possible the registration of a particular substance. This occurs through electronic data processing and data transfer. To be sure, all these developments make possible scientific work of increased scope and complexity. However, they also make research more expensive. Furthermore, the inflation rate for research and development is typically fifty percent above the overall rate of inflation.[2] To put everything over a common denominator: In relationship to other social services, the expenditures for health care are too high. On the other hand, demands on the health-care system are also rising. To be sure, drugs make up only a part of these costs (in Germany about twenty percent, in the United States nine percent, in the rest of Europe somewhere between these two figures),[3] but they are greatly affected by the measures taken to reduce health-care costs. The therapeutic value of a drug is measured by the cost of the disease to be treated and by the cost of other medicines or treatment modalities for the same indication. On account of high development costs and price cuts, medicines that offer nothing new in relation to already existing treatments have no chance of economic success. Therefore, drug redundancy will decrease, and consequently, a source of income for the industry will be lost. Put more simply, the pharmaceutical industry must accomplish more than it has to date with more modest financial resources.

## Strategic Options

What options are at the industry's disposal? Fundamentally, there are only two: an offensive and a defensive strategy. The defensive strategy is to compensate for smaller profit margins through higher sales volumes. To go this route it is important to attempt to control the distribution of the product. In Europe this is less of an issue than in the United States, where a number of buyers can unite in health maintenance organizations (HMOs) to force price competition. To some extent this is already succeeding. An HMO is a profit-oriented business that tries to offer health-care services to its "customers," that is, to as large a group of patients as possible, at the best possible prices. Such a strategy is successful when an HMO represents a large number of individuals (ideally several million) and when it is thus in a position to negotiate fees with hospitals, group practices, diagnostic firms, and other providers of diagnostic or therapeutic services, by whom patients (insured or not) will be cared for at acceptable prices, while at the same time the HMO is able to make a reasonable profit. The HMO compels hospitals and physicians to offer their services at a discount and makes its profit on the difference between what the patients or their insurance pays them and what they themselves pay to the clinics and physicians for their services. In the process, the HMOs often determine what medicines will be prescribed in routine cases and what medicines will never be prescribed or are permitted to be prescribed only in exceptional cases. They determine, often in consultation with physicians, what treatments are to be applied in particular cases, how long a patient will be allowed to remain in the hospital after a particular operation, and similar such restrictions. Thus HMOs are organizations that attempt to turn medical treatment into an economically viable process. However, in the process they involve themselves in medical decisions in a manner subject to increasing criticism, not always to the benefit of the patient. There is hardly any dispute that the "managed care" practiced by HMOs has led to an economizing of medical services. But one may at least doubt whether they are consistently contributing to a better practice of medicine, one more suited to the needs of the patient. If the estimates are correct that the American market will be seventy to eighty percent controlled by HMOs by the year 2000, then inevitably certain constraints will result. Various large pharmaceutical companies have concluded alliances with HMOs to ensure that their medicines will have the widest possible distribution among the population served by an HMO. The acquisition of MEDCO by Merck for six billion dollars, which shook the industry in 1993, is an example of this trend. What distribution companies and HMOs offer in addition to the preferred prescription of particular drugs are, of course, data on patients, on the prescribing habits of physicians, calculations of economic viability, that is, pharmacoeconomic data, all of which is information that is useful in positioning a drug optimally in the market. This has nothing to do with an increase in the quality of medical care. In the foreground is inexpensive medical care, not medical care offering the best possible diagnosis and treatment.[4] It is perhaps understandable now why this strategy, based on *volume* and *distribution*, can be described as "defensive."

On the other hand, an offensive strategy rests on the conviction that new types of medicines that can prolong life, improve the quality of life, and do this to a considerably greater degree than could earlier drugs will also have value in the future. Whoever is capable of offering good—that is, medically and economically viable—medicines at regular intervals (about once or twice a year) should be able to achieve success even in the restrictive markets of today and tomorrow. Strategies that rest on this conviction may be called "offensive." We are also speaking of innovative strategies. However, innovations depend on new or innovative products, and new products can be obtained only through research and development. And that costs money. The pursuit of a strategy of innovation in today's economic and technological environment can mean only that one wishes to deal more productively than before with the resources allocated to research and development. For this there are two paths: the improvement of one's own research, resulting in greater productivity, and collaborative work with third parties. Before we discuss in detail these options and their relative weights in a strategy of innovation, we must learn in what ways industry is capable of discovering and developing innovative medicines.

## An Innovation Deficit in the Pharmaceutical Industry

A study carried out in 1995 set out from the idea that the number of new products that can be brought to market by one or more firms should correlate reliably with the number of preclinical research projects carried out five to six years before the intended introduction of the product.[5] The authors of this study thus determined the number of preclinical research projects that were carried out in the pharmaceutical industry in 1993. In several cases they were able to rely on direct information, while in most cases they had to estimate the number of projects on the basis of published research costs. This was possible because in the case of several firms the number of preclinical research projects as well as the research costs were known. It turned out that in 1993 there were 1,300 preclinical projects carried out by fifty firms worldwide. Corresponding numbers were obtained for the ten leading firms as well as for the top twenty or thirty pharmaceutical firms. Considering the historically known rates of success for the transition from research products into development and from development products to market, they could estimate how many commercially viable drugs would result from the projects undertaken by the industry in 1993. To be sure, this number had to be divided by the number of years necessary for the complete renewal of a research portfolio. Here the authors used a historical number calculated on the basis of real examples that covered a period of four years. They took six years as the development period and then calculated the number of products to be introduced in 1999. This calculation is based on the hypothesis that the number of drugs making the transition from research into development is approximately the same in each of the four years required for the renewal of the portfolio. It is worth noting that the expected number of approvals, whether for the industry as a whole (all fifty firms) or for groups of the top ten, twenty, or thirty firms was lower than the corresponding figures from the recent past. In other words, the trend that has been observed since 1985 of fewer product introductions seems to be continuing.

If we compare these sobering findings to the industry's expectations for growth, there appear to be considerable discrepancies, depending on the expected rates of growth, between what the industry *can* deliver and the growth that it *wants* to achieve. To estimate roughly this discrepancy, assumptions must be made about the life expectancy, the average annual sales, and the length of the development process of a "standard drug." The authors decided on an average sales volume of $400 million per year and assumed the average life expectancy of new drugs to be seventeen years. The length of the development time (from the beginning of development until the first registration) was taken as six years. This comparison showed that with the projected number of registrations, the industry would not be capable of maintaining its current size: even with zero growth there was a deficit of eleven new substances per year. According to this calculation, the ten leading firms would have five substances too few; for the leading twenty the innovation deficit stood at seven new substances per year. With an average five-percent growth of the industry, the deficit would stand at nineteen substances per year, while with a ten-percent rate it would be thirty substances per year (Table 7.1).

Based on these figures, it is unlikely that the industry as a whole can produce any significant growth. This does not exclude the possibility that individual firms will continue to be successful in the future, based on good research accomplishments. For the industry at large, however, these numbers underscore the necessity for bringing their size into accord with their innovative basis if the negative development described above is to be averted.

## Increasing Research Productivity Through Mergers and Acquisitions

One obvious way of increasing the output of a pharmaceutical company is to acquire skills, functions, and products through mergers and acquisitions. In recent times many such events have taken place.

I shall list a number of strategic reasons that may have been at the heart of a selected number of recent mergers and acquisitions in the past. The key word in this context is "complementarity," be it scientific, techni-

Table 7.1.   Innovation deficit in the leading pharmaceutical firms.

| Growth rate | For all firms | | | Per firm | | |
|---|---|---|---|---|---|---|
| | Top 10 | Top 20 | Top 50 | Top 10 | Top 20 | Top 50 |
| −15% | 1.9 | 2.9 | 3.9 | 0.2 | 0.2 | 0.1 |
| −10% | 0.2 | 0.4 | 0.2 | 0.0 | 0.0 | 0.0 |
| −5% | −1.9 | −2.9 | −4.6 | −0.2 | −0.2 | −0.1 |
| 0% | −4.7 | −7.3 | −10.9 | −0.5 | −0.4 | −0.2 |
| 5% | −8.2 | −12.8 | −19.0 | −0.8 | −0.6 | −0.4 |
| 10% | −12.7 | −19.8 | −29.2 | −1.3 | −1.0 | −0.6 |
| 15% | −18.4 | −28.6 | −42.1 | −1.8 | −1.4 | −0.8 |
| 20% | −25.5 | −39.6 | −58.1 | −2.6 | −2.0 | −1.2 |

• The numbers in the two right-hand segments denote the difference between the numbers of new introductions that seem within reach and the number that would be necessary to attain the rate of growth given in the left-hand column.
• At a growth rate of −10%, there would be no deficit according to these figures.
• At a growth rate of zero, there would be an annual deficit of eleven substances; at 10% growth, it would be thirty substances per year. The numbers were computed for the leading 10, 20, and 50 firms on the basis of data from 1993.

cal, related to products, or geographic. The acquisition of a majority of Genentech by Roche in 1990 marked a conscious commitment on the part of the acquiring company to buy into molecular biology on a large scale and in particular to exploit the promise of recombinant proteins, monoclonal antibodies, and other therapies that would emerge from molecular biology. The acquisition of Syntex by Roche a few years later (1994) was very much dominated by the desire to achieve a strong geographical balance: Syntex had a strong presence in North America and in Mexico, areas in which Roche needed more strength if it was to become a top player in pharmaceuticals. Of course, product complementarity played a role in both cases with respect to marketed products as well as to products under development. This point will be revisited later. In the case of Roche and Syntex, the novel immunosuppressive drug CellCept represented an incentive for the deal as did Ganciclovir, a potent antiviral with a special use in cytomegalovirus infections that occur as a complication to AIDS. At the time, Roche was developing novel drugs for both indications, immunosuppression related to organ transplantation as well as to autoimmune diseases and AIDS. Tenefuse, a recombinant antibody construct, and the first protease inhibitor were in the making—Cellcept and Ganciclovir were welcome and perfectly complementary additions to these development drugs.

Both the geographic and the product-related complementarities have led to the expected synergies. Especially in the case of Tenefuse and Cell-Cept this therapeutic and commercial synergy is only at its beginning and is by no means exhausted.

In a similar vein, one could argue that the merger between Glaxo and Burroughs Wellcome to create the then biggest pharmaceutical company was at least in large part driven by geographic and by product complementarities. Burroughs Wellcome had a rich portfolio of antiviral substances and the know-how to generate more of these compounds, a field that Glaxo needed to strengthen. Also, Wellcome had been known for its work in basic biology, which appeared to many as a necessary complement to Glaxo's strength in pharmacology and medicinal chemistry. Product and/or geographic complementarities were also obvious factors in the merger between SmithKline and Beecham as well as that between Squibb and Bristol-Myers.

But what about the most recent event, the impending merger between Glaxo Wellcome and SmithKline and Beecham? This looks like an attempt to build a perfect pharmaceutical company, perfect at least by today's standards and expectations. Glaxo had not been a fast assimilator of biotechnology when recombinant DNA products and monoclonal antibodies were beginning to emerge. Its purchase of the former Biogen, in Geneva, to create a Glaxo Institute of Molecular Biology represented a rather belated response to the challenges and opportunities introduced by molecular biology. While the company has since shown a rather keen interest in molecular biology, especially in genetics, it does not have in this area the broad-based experience and database that SmithKline and Beecham started to assemble in 1993 through its association with Human Genome Sciences. So from a strategic point of view one could argue that Glaxo Wellcome's strength in medicinal chemistry, pharmacology, animal biology, and its recent buildup of modern screening technologies would fit well with SKB's expertise in genomic sciences.

## The Question of Critical Size

A company must be able to generate novel compounds through its own R&D, through alliances, and through licensing activities at a rate that will support healthy growth. It must also be able to utilize its drugs globally and efficiently in all major markets. Today, the revenue size necessary to sustain such activities may be in the range of 6–8 billion dollars. Companies that lie significantly below this range will hardly be able to compete internationally in a broad array of fields. They will either disappear as independent entities or convert themselves into highly specialized "niche players," either topically, geographically, or both.

On the other hand, it is by no means certain that companies that operate significantly above this range are likely to be more successful than their smaller peers. At least in the past, productivity beyond the critical size of 6–8 billion in revenues did not necessarily correspond to size. To operate on a very high revenue base will force a company to generate at least one "big" drug per year, possibly more. So far, however, the capacity to generate "blockbusters" (drugs commanding sales of more than one billion dollars) has not been a function of size or money spent for research. There is a possibility that new technologies will change this. Methodological completeness, if well managed, may make a critical difference. However, there is no proof of this.

The combination of many targets (derived from genomic research) with combinatorial chemistry and high-throughput screening will produce a large number of "hits." But those hits may not represent the best ways of correcting a critical physiological parameter. While the new technologies will allow the identification of a great number of potential targets and substances that "bind" to these targets, cell biology and physiology will still be needed to identify the most appropriate sites of pharmacological intervention. In the end, good judgment based on integration of many details into realistic functional images of physiological or pathophysiological events will still be needed. Eventually, computer simulations may accelerate such target validations. At present, however, there is no replacement for the integrative capacity of biologists and physicians. The positive impact of the new technologies on research productivity may therefore not be as quick as some managers and observers of the industry seem to expect.

Size is also important in the context of the marketing of a drug. The complexities of research and development as well as of drug regulation have made the delivery of a new drug an extremely expensive endeavor. During the last 20 years the costs for finding and developing a new drug have risen by a factor of ten. In constant dollars they have quadrupled. Such expenses can be recouped and turned into profits only if a company can market its products on a worldwide scale. If a global presence has not been achieved historically, it must be generated by acquisitions or by merger. This global marketing presence must be strong for competitive reasons. In an ideal world, a novel drug addressing a compelling medical need in a new way should not need much marketing support other than expert guidance to the medical community (and perhaps to patients) on how to best use the drug. In the real world, however, innovations tend to occur in clusters. Therefore, products need to be differentiated against similar products in a competitive environment. In order to achieve this, a company must have not only a global presence but a *strong* global presence. A critical size is the key to both, and often size can be achieved only by mergers or acquisitions.

### Broader Strategic Issues

Mergers like the Merck–Medco deal and similar acquisitions of HMOs by other big pharmaceutical companies or the recent acquisition of Boehringer Mannheim by Roche are driven by broader strategic considerations: In markets that have become increasingly competitive, it appeared useful, if not mandatory, to enhance or at least secure the distribution and utilization of one's own drugs through the specialized organizations that provide health care, including the provision of drugs to millions of patients, and that could give pharmaceutical companies insight into the relevant patient database. The Roche–Boehringer Mannheim merger was driven by the desire to make Roche's diagnostic business a powerful, technologically complete and globally acting division as well as by a long-term expectation that in the emerging healthcare markets of the future companies that will be able to provide broad health-care packages including drugs and diagnostic tools will have an advantage over companies that can only make and sell drugs.

Complementarities within the confines of the pharmaceutical business per se, broader strategic issues related to changing health-care markets, and critical size as the key to a global presence in marketing and manufacturing: These are important reasons for mergers and acquisitions, and to a remarkable extent, past acquisition and merger events have fulfilled these expectations.

### The Issue of Productivity

There is, however, one additional and very serious factor that drives mergers and acquisitions. It is not always easy to recognize, but it reaches even deeper into the soul of an enterprise than any of the factors mentioned above. We are talking of productivity. Almost invariably, the resolve to merge or to acquire is influenced by an imbalance between the cost of R&D and that of other operations on the one hand and revenues on the other. When companies feel that their present product portfolio and their portfolio of development drugs will not suffice to sustain their expectations for future growth, they tend to acquire additional products through mergers or acquisitions and to build a combined portfolio that will then give them more profitability in the near and intermediate term. In order to derive maximal profits from an expanded portfolio, companies after mergers must reduce their costs and—if possible—at the same time improve their operations. "Downsizing" has become a popular term for this operational necessity. After a merger or a major acquisition, many positions become redundant. Suddenly, there are two candidates for only one position. This does not apply throughout the organization but through large parts of it. Obviously, it would be logical to retain the stronger candidates and to separate from the weaker ones. Unfortunately, there are limits to such rigorous choices: Time is one serious limitation. Consolidations are usually carried out under time pressure. Judgment, especially when distorted by political and cultural influences, is another one. Factors related to the compatibility of functions and people should influence choices to be made but are, of course, sources of errors. Social considerations add to the list—especially when the scene of a merger or acquisition involves European countries. Sometimes the choice of managers or scientists to stay on in the new company are based on complementarity considerations more than on an objective assessment of professional competence. For all of these reasons, the organization evolving after a merger may not be significantly stronger than the two parent organizations. Such a result could, of course, compromise the merger or acquisition in a serious way.

### Securing Long-Term Productivity

The most serious mistake that is made during mergers relates to research and development. Here it is particularly difficult to select the right individuals for the future company, because the review that leads to these critical personnel decisions is often biased. In all cases, it will be dominated by the stronger partner. The consolidation of projects and people can under such circumstances not be a "peer review" in the objective scientific sense.

However, there is another consideration to be made. If a company does not possess the product line and/or the development portfolio needed for growth, these deficiencies often have their origin in the productivity of the R&D departments. An imbalance between productivity on the one hand and growth expectations on the other is among the reasons that propel mergers and acquisitions in the first place. If this simple fact would be taken into account, managers would be extremely cautious in downsizing R&D organizations after mergers or acquisitions. Instead, they would intensely concentrate on building stronger and more productive organizations than either of the two parent groups had been. This would take time and would in many cases result in only moderate downsizing. The emphasis would lie on an increase in quality and output. The resulting R&D organizations would perhaps be somewhat smaller than the sum of the parent groups, but they would have to be better in the first place. As a proportion of total revenues of the resulting company, the R&D group might even be larger than either of the two parent organizations had been in relation to the revenues of their respective parent companies. The purpose of managing the R&D merger very carefully and placing priorities on the improvement of quality and productivity rather than on cost savings would address the long-term need for sustaining increased revenues rather than on just creating them by adding 2 and 2 together.

In reality, this careful adjustment of research size and productivity to the situation resulting from a merger is almost never done. Instead, R&D departments are reduced according to parameters that are largely numerical in nature. Functional aspects are often not understood and not addressed with sufficient care. Therefore, in most mergers and acquisitions some R&D organizations are hit twice: in the first place by sudden downsizing measures, which are always demoralizing even to those who will stay on, no matter how tactfully and humanely they are carried out. The fact that good style is not a trademark for the pharmaceutical industry as a whole in this respect, especially not for American companies, shall be mentioned only as a footnote. Second, however, R&D after a rigorous downsizing may not be strong enough to generate a faster flow of products, which is now needed in order to sustain a larger product portfolio and an adequate growth rate for the higher revenue base that resulted from the merger. In other words, in most mergers and acquisitions the emphasis is on short-term optimization. Insufficient care, if any, is applied to address the underlying issue of productivity that motivated the merger in the first place. Therefore, after relief, which may last for a number of years, the old imbalance between research productivity and growth expectations may again become obvious and lead to a new round of mergers and acquisitions.

In the final analysis, of course, such a "merry go-round" will not work. Research will not thrive in an environment that is governed mainly by schematic financial considerations for the short term. Talented young scientists who are interested in the numerous applications of modern biology and chemistry in medicine have alternatives. Even in European countries they often have the chance to establish and build their own firms or to join a growing technology- and discovery-oriented industry in the United States and in some European countries.

The present consolidation activities in the pharmaceutical industry may therefore lead to vicious cycles: more acquisitions and mergers in the not-too-distant future because the underlying evil, the lack of productivity of R&D, was not adequately addressed in the first round. If not carried out with a better understanding for the conditions under which research productivity can be enhanced, these waves of mergers and acquisitions may lead to an even more radical consequence: to an exodus of discovery research from big pharmaceutical companies and to the establishment of a universe of discovery companies on whose services the big pharmaceutical companies will increasingly depend.

## Increasing Research Productivity Through "Internal" Measures

Innovations, then, are a "scarce commodity."[6] During recent years, the focus of all measures to promote productivity lay unequivocally in development. It was clear that here, often years were wasted, and with a product with sales of several hundred million per year, several years' delay in development represents a loss of sales whose size is a function of the product of the annual sales and the length of the delay. The leading pharmaceutical firms are constantly attempting to improve their development performance. We may expect that they will optimize this sector of providing drugs within the next few years, so that perhaps in five to ten years the development process will have been so thoroughly mastered that it will no longer have much importance as a distinguishing competitive characteristic. What will then determine the success or failure of pharmaceutical firms will be their capacity to *find* new products to develop. The real difference between successful and unsuccessful firms could lie, then, in the quality and quantity of innovative products that will be made available each year for development. Herein lies the deficit that we are observing today. Here, then, is the point of departure for corrective measures.

## More Productive Research Units Through Greater Autonomy

It is a question of increasing the productivity of research. This can be accomplished through internal measures or through collaboration with third parties. When we consider internal measures, we need to consider that the large firms, with their ponderous and bottom-line-oriented organizational structures, do not offer an ideal environment in which research can flourish. Since research cannot produce optimal results in such an environment, it should consciously be granted a measure of freedom that it now enjoys in only a very few firms. There are indications as to how this should be accomplished: We know of hundreds of small biotechnology firms that considering their size are extraordinarily productive. Productivity in this context does not mean profitability. Rather, we use the term to say that a small number of employees have been able within short time periods to achieve remarkable results with relatively small expenditures. The research productivity in the larger firms could certainly be improved if the large research structures were to be divided into smaller "semiautonomous" units and if these units were to be granted a higher degree of entrepreneurial and scientific freedom. An interesting model for the future could then consist in dividing the exploratory research of a large firm into several thematically defined institutes or research firms, each with its own budget and each to be evaluated on the basis of its individual contribution, while granting to them the freedom to determine for themselves by which methods success should be achieved.[7]

During the last few decades, pharmaceutical companies have, if anything, tended toward a stronger integration of research and development into other corporate functions. For development this makes sense, since in working out or confirming a broadly accepted profile of efficacy, problems of marketing are immediately addressed. Similarly, regulatory decisions have to be taken together with the clinical research organization, and production methods must be worked out for an adequate drug supply in support of late clinical studies and for marketing. All of these things need careful coordination. Development is a multifunctional process, which requires precise planning, good organization, and considerable discipline.

However, the integration of research into other functions of the company has many negative consequences. Research requires room to maneuver. Its results cannot really be planned for. Its course can thus not be determined by desired results, but must be driven by fundamental questions and methods. This does not mean that research should be uncoupled from the rest of the enterprise. On the contrary: Research management must help to develop and support the company's overall strategy. It must also transmit the relevant requirements of such a strategy to the members of the research organization. Quantitative and qualitative goals of a research center should be formulated together with the firm's management. The budget worked out between research management and the firm's overall management must be commensurate with the global strategic framework and operational goals. Yet within this thematic and financial framework research must be free to create its own opportunities. The choice of scientific methods, collaboration with other scientists, the organization of the center, the recruitment of employees—all this should be left to the research center itself to determine. It is easy to list the criteria by which a research center should be evaluated: reaching its goals (quality and quantity of substances that go into development over short- and mid-range time periods of one to five years), patents, publications, staying within budget, avoidance of duplication with other centers or conversely the creation of synergistic effects, and participation in the implementation of the firm's overall strategy through greater collaboration with biotechnology firms or other partners.

## Quality Begets Quality

There is no general recipe for the establishment of a research center. It is a good idea, nevertheless, to have a look at particularly successful institutes or scientific establishments in order perhaps to note some of the factors frequently associated with success. The most important prerequisite for success is having competent, motivated scientists. We might use the expression "critical quality." Quality begets quality. Good scientists attract other good scientists. Incompetent or even mediocre researchers have the opposite effect, especially when they are in supervisory positions. Whoever would have productive centers must have the courage to give a great deal of freedom to first-rate employees and to jettison weak ones. Additional characteristics of good centers are sufficient size and proximity to other scientific establishments. More than ever, science is an interactive process. An institute or center that simply stews in its own juices and has to make do without stimulating daily contact will not be productive. The optimal size for a pharmaceutical research center is several hundred employees. This number should include representatives of all disciplines necessary for the early characterization of drugs, therefore toxicologists, biochemists, pharmacologists, and clinicians. Centers with more than six to eight hundred employees tend to anonymity. With fewer than two hundred, a center is in danger of becoming methodologically and conceptually too narrow. Good scientists are generally prepared to accept authority based in a knowledge of their discipline. To a research organization, seniority or structural principles taken from other walks of life, such as the military or governmental bureaucracy, are poison. Nonhierarchical organization, informality, openness, and a congruence between scientific capability and personal integrity on the one hand and responsibility on the other are typical indicators of successful research organizations. Financial incentives are helpful when they are based on performance.

## Where Does Motivation Come From?

Bureaucracy, electronic recording of hours worked, operative restrictions by supervisors, long and tiring meetings with primarily administrative content—these are surely, and unfortunately they are typical, methods in the industry for decreasing productivity and discouraging talented and motivated employees. Even more dangerous than these formal obstructions are substantive restrictions, for example setting researchers' goals based on short-term marketing requirements. Such phenomena are ominous and signal serious problems for research. In the pharmaceutical industry, especially in research and development, it has become common to emphasize motivation of researchers as a critical attribute. In real life, however, truly creative researchers are, like other creative people, motivated *a priori*. Motivation is part of their life force, which is directed towards self-selected goals. Conditions can—and should—be provided under which this energy can flourish, and negative influences can be avoided as well. To believe that uninspired, opportunistic employees can be motivated to creative achievement is a childish error whose origin can be found in a simplistic behaviorism that has been abandoned even in the United States. This experimental subdiscipline of psychology was based on the assumption that the behavior of animals (and human beings) in response to particular stimuli could be observed and shaped under controlled laboratory conditions. Modern behavioral research, observing innate behavior in its natural milieu, has shown that behaviorism's approach failed to consider important, namely inherited, components of behavior. Therefore, if one has problems with the creativity of one's employees, one has probably hired the wrong individuals. Motivational contortions will prove futile.[8]

In sum, the key to increased research productivity rests in the recruitment of first-class, creative, and motivated scientists. Such employees find satisfactory working conditions ever more rarely in the classical pharmaceutical industry. This is why we advocate the establishment of autonomous research centers that operate in a strategic and financial framework that is defined and implemented together with company management. Only when it understands that for maximal productivity, research requires conditions that distinguish it from the conventional corporate structure will the industry be able to regain its central role in drug research. The alternative is the loss of research as a typical function of the pharmaceutical industry, with a resulting limitation to development and distribution.

## Collaboration with Others

Collaboration of pharmaceutical companies with small firms, in particular with biotechnology companies, is well established today. Such collaboration can accomplish a number of goals. In the past, the most frequent type of agreement was an alliance formed around a particular technology that could assist in drug discovery (fifty percent of all such agreements). The second-largest component of agreements, about thirty percent, concerned the joint development of already selected products. The remaining twenty percent dealt with purely methodologically oriented collaborations or with agreements that cannot be more precisely classified. In 1994 there were 117 formal contracts drawn up between biotechnology firms and pharmaceutical firms. By 1995 there were more than 140 (Burril und Craves, *Bioworld* 1995).[9] In 1997 180 such agreements were concluded. The principal areas of interest of the pharmaceutical firms taking part in such agreements were the chemistry of small molecules, gene therapy, genome research, intracellular regulation, mechanisms of aging, novel mechanisms for suppressing cell division, new drug delivery systems, and combinatorial chemistry. The biotechnology industry, which we shall discuss in greater detail below, has become an important partner of the pharmaceutical industry. It has greatly enriched the research palette of the pharmaceutical industry through the production of new ideas, new methods, and innovative research techniques. But it is also making its effect noticeably felt in the very tangible realm of newly approved products. As of February 1997 more than forty monoclonal antibodies and recombinant proteins had been approved worldwide as drugs. The proportion of "biotechnologically" obtained drugs out of all new drugs is growing constantly and was at about twelve percent in 1995. In 1998, fifteen biotech drugs have been or will be approved. By the end of the decade, biotechnologically obtained drugs will constitute just under one-half of all newly introduced medicines. The first wave of the biotechnology revolution has indeed enriched the drug market (see Table 7.2), and the pharmaceutical industry is making good use of the scientific and technological potential offered by biotechnology firms. Thus the productivity of this industry can also help to reduce the innovation deficit in the pharmaceutical industry.[10]

With the universities it is quite another matter. Here the great potential for collaboration appears by and large to be underutilized. Of course, universities have traditionally been scientific partners of industry, but this partnership reflects more the traditional roles of the two institutions than

Table 7.2.   The methods of molecular biology are introduced into basic research, in the search for new drugs, in diagnosis, and in the production of genetically engineered medicines.

**Introduction of Gene-Technological Methods in the Pharmaceutical Industry**

| Gene-recombinant products | Diagnosis | Drug research and development | Basic research for treating the root causes of diseases |
|---|---|---|---|

Examples

| Human insulin Growth hormone Erythropoietin (EPO) Interferons Interleukins Factor VIII/IX t-PA Hepatitis B vaccine Humanized monoclonal antibodies Fusion proteins | Monoclonal antibodies for diagnosis of infections, cancer, etc. PCR (polymerase chain reaction for qualitative/quantitative determination of HIV/HCV, etc. Tumor diagnosis Genetic tests | Tracking down new body-specific proteins (TPO, leptin) Isolation and reproduction of "drug targets" Improvement of existing drugs through clarifying the mode of action | Decoding the human and bacterial genomes (Genome Project) Transgenic animal models "knock-out" mouse |

Consequences

| • High degree of drug safety • High specificity • Standardized quality • High economic viability | • Early diagnosis • Goal-directed treatment • Disease management • Predictive diagnostics | • Goal-directed screening | • Somatic gene therapy • Influence on gene expression/ protein synthesis |

their possible complementary roles in the creation of innovative products and technological solutions. In the past, universities frequently received stipends, research support, gifts, financing of facilities, and the like from industry. In the United States, professorships were often endowed by industry. These contributions were generally made with no strings attached, with the tacit assumption that university research would one day prove itself relevant to industry and that industry would always be needing a supply of well-educated scientists. By and large, these assumptions turned out to be correct. Over the past twenty years, however, there has been a change, for which several reasons may be cited. The interest of university researchers in the practical application of their research findings in the biomedical area has grown, particularly in the United States. On the one hand this has come about because molecular biology, and thereby also the related biological sciences of immunology, cellular biology, neurobiology, and others, had reached a degree of methodological maturity that greatly encouraged applications to medicine. For example, in the 1950s and 1960s, bacteria and bacteriophages were the principal model in molecular biology. Later came yeasts, threadworms (*Caenorhabditis elegans*), and the fruit flies *Drosophila*, which had been of interest to geneticists for decades. In the 1970s and 1980s research extended to eukaryotic cells and later to mammals, particularly the mouse. Today, a central concern of molecular biology is the decoding of the structure of the human genome, the functional understanding of genes and their relative contribution to physiological research and to normal and pathological development. This means that new, exemplary discoveries that have been obtained from basic research do not have to be *translated* into a connection with medicine: They are *already in* such a connection.

There is perhaps a further, culturally based, reason for an increased interest away from basic research in favor of practical applications of scientific knowledge. As Erich Kästner put it, *"Es gibt nicht Gutes, ausser man tut es"* ("There is nothing of value except what is *done*"). The often repeated doubt whether past investment in basic research has led to an improvement in the human condition is met with a call for practical solutions to problems. From such a viewpoint this change of direction by many basic researchers toward applied research, especially to the solution of problems of diagnosis and treatment, would represent a most effective method of combating a widespread skepticism regarding the value of science. This skepticism, together with growing demands from other quarters for public funds, has led in recent years to a stagnation—indeed, a reduction—of public funding for research and development. This is true particularly for the United States, but it applies to other countries as well.[11] Universities have been compelled more than previously to seek money for research from sources other than public funding agencies. A demonstration to a broad segment of the population of the effectiveness of the translation of fundamental scientific knowledge into practical measures for the improvement of society could be effective over the long term in convincing governments and the general population that investment in science is worthwhile.

## Breaking Out of the Ivory Tower

All of these factors have contributed to the result that leading universities, above all those in the Bay Area of San Francisco, have become breeding grounds for new biotechnology firms. Of course, there are other factors at work, including a love of adventure and the desire to get rich quick. Yet without the above-mentioned factors the biotechnology industry would not have arisen out of the universities so rapidly. Culturally, the leap from a university position to a small biotechnology firm certainly represents a change, but not a drastic alteration of conditions of life and work. American universities quickly learned to adapt themselves to the new situation. In the 1980s they had already begun to grant time to faculty members who wished to found or consult to biotechnology firms (generally one day per week) and to institute rules for faculty compensation in relation to such activities. The interplay between the biotechnology industry and American universities functions well today in general, although there are regional differences to be noted. On the other hand, the direct collaboration between universities and the pharmaceutical industry in still in its infancy. A recently published survey shows that there are six thousand collaborative projects underway between biomedically oriented firms and universities, of which most last about two years and cost less than one hundred thousand dollars.[12] Thus the relationship between industry and the university is characterized by a number of relatively small-scale interactions of short duration. About thirty percent of these are clinical projects, carried out in university facilities under contract with industry. There are also frequent consulting contracts with individual professors as well as support of small research projects. An opinion poll among scientists in industry and the university indicated where the difficulties lie. University researchers welcome collaboration with research establishments for three reasons. First, they want to be part of the transfer of technology. The motto, "From bench to bedside," is very attractive to them. Second, they obtain access to new substances and technologies. Finally, they can finance part of their own research with money from industry. They experience as disadvantages possible conflicts of interest and the fact that an authority from outside the university eventually will have an influence over the course of their research. Conversely, scientists in industry were interested in collaboration with universities mainly to obtain access to new ideas and methods. Furthermore, they hoped thereby to recruit young researchers from the university into their organizations. They experienced as obstacles the bureaucracy of universities, the loss of control over joint projects, and the difficulty that academics have in dealing with confidential information.

In the research-based American universities in particular, there is an enormous, largely untapped potential for the exploitation of practical applied concepts and methods. At present a process of rethinking is taking place in industry and universities that could ease the way for broader collaborations in the future. We shall examine this issue more closely.[13]

## Focus of Innovation: The Biotechnology Industry

The American biotechnology industry is the unforeseen consequence of the revolution in molecular biology that began in 1944 with the discovery that the DNA molecule is the carrier of genetic information. In 1953 James Watson and Francis Crick elucidated the structure of DNA. There followed in rapid succession a delineation of the important steps in the DNA–RNA–protein synthesis process. The genetic code—the language of genes—was deciphered, and soon thereafter the enzymatic instruments of DNA recombination were isolated. Now their function could be understood. In 1973 the first expression of an animal gene in a bacterial cell was accomplished, and two years later Georges Köhler and Cesar Milstein discovered the principle of monoclonal antibodies. In 1976 the first gene-technology firm, Genentech, was established, and a number of similar firms soon appeared. Today (1997) the American biotechnology industry comprises more than thirteen hundred firms, of which eight hundred have a therapeutic orientation. In 1994 these firms together achieved sales of $11.2 billion. Their costs during this same period totaled $15.3 billion. Thus despite the success of some companies, the biotechnology industry as a whole is still running a deficit. Nonetheless, in 1994 the industry employed 103,000 persons. In addition to the therapeutically oriented companies there are many diagnostic firms, with agricultural enterprises in third place. Environmentally oriented firms and producers of biochemical reagents play a relatively small role. The biotechnology industry today is capitalized at more than $50 billion. As of 1997, in the United States there were 485 small firms (1 to 50 employees), 422 medium-sized firms (51 to 135 employees), 236 "large" firms (136 to 299 employees), and 147 "top-tier" companies with a work force of over three hundred. Ten "breakaway" firms, among which are Genentech, Amgen, Biogen, and Chiron, have one to several thousand employees, run international operations, and have grown in their particular industry into fully integrated firms.[14]

Europe, too, has meanwhile developed a biotechnology industry. It comprises just under five hundred firms, none of which has more than three hundred employees. Moreover, the various divisions of therapeutics, diagnostics, agriculture, instrumentation, and environment are roughly equal in size.[15]

The therapeutically oriented, that is, the "pharmaceutical," biotechnology industry has meanwhile brought forth a considerable arsenal of medicines. As of today (1997), more than forty recombinant proteins and monoclonal antibodies have been approved, and another 450 are in one stage or another of clinical development, of which 127 are already in Phase III. Many of these medicines have only a local significance, and the large numbers indicate that we are dealing with a considerable degree of redundancy. Yet we may assume that the annual number of new products coming out of today's biotechnology industry will grow in the coming years from the current six or seven to something like ten to twenty. This production can be largely attributed to the large number of recently established publicly traded companies (about three hundred such firms in the United States in 1996). Thus the biotechnology industry is acting as the industrial incubator for new technologies. In particular, genome research, combinatorial chemistry, but also gene therapy, whose home turf is primarily in the medical centers of large universities, are strongly represented in biotechnology companies. Chemical technologies, too, are leading to the establishment of new enterprises. Examples are firms such as Arris and Ariad. Thus the biotechnology industry may well develop over the next ten to fifteen years into a broad-based research industry that provides substances for development to a pharmaceutical industry that has meanwhile become specialized in development and distribution. There is much evidence that such a division between research and development will, in fact, occur. We shall discuss this issue in a later section.

## Centers of Innovation in the Universities

We have seen that American universities whose biological and biomedical research have been generously supported since the end of the Second World War by the National Institutes of Health became the breeding ground for an entirely new industry. Even today the universities maintain strong ties with this industry. The transfer of technology via university researchers from the university to small, newly established enterprises is a continuing process. In absolute numbers, fundamental research in biology in the United States is relatively well supported, with more than $12 billion per year. But the amount of investment is growing very slowly. Taking a rate of inflation of about five percent (for scientific research inflation runs at about fifty percent above the overall inflation rate), this stagnation represents a de facto decrease in the support of basic research. Younger researchers are affected most of all. If established laboratories are to be supported in the future only at current levels, relatively little money will be available to younger scientists in the process of establishing their careers. This has developed into quite a precarious situation, since the number of doctorates awarded in the natural sciences has been growing. An additional factor, one that falls most heavily on the medical schools, comes as a result of the restructuring of the American system of health care, which is controlled more and more by HMOs, which represent millions of patients and are able to compel hospitals to compete with one another for the right to care for an HMO's patients. In this competition research hospitals find themselves at a disadvantage in relation to those that focus entirely on patient care, since part of their budgets are devoted to research. Thus university institutes and research clinics must begin to seek additional sources of funding more vigorously. In this regard industry is an obvious potential partner, though of course, the cultural differences that we mentioned earlier stand in the way. But the pharmaceutical firms as well, at least the traditionally research-oriented among them, must, in view of a threatening deficit of new relationships, show an interest in broadening their innovative basis. Universities and pharmaceutical firms have complementary interests, and this should encourage them both to seek new models of collaboration. Such models must ensure that the fundamental interests of both institutions—pharmaceutical industry and the university—are kept in focus and that both are able to improve their situations. How might this be accomplished?

Certainly, an extensive and unconditional support of basic research by industry is no longer viable, as we have already mentioned. Collaborative projects must be structured so that the financing party, the particular firm, receives within a reasonable period of time what it requires: critical scientific information, concepts, experimental models, methods, and in some cases substances for development. It is necessary that these conditions not only be accepted by the academic partner, but also that they be actively promoted. On the other hand, it is important that the university receive something as well and is not just the executor of industry's intentions. Those scientists from industry responsible for the collaborative project or directly involved in it must encourage in the universities the growth of fundamental research, even if the results of such support are not immediately relevant for medical practice or the particular interests of the firm. Both sets of interests—the concrete expectation of experimental progress as a prerequisite for a therapeutic research program and the desire for fundamental knowledge—must be reconciled. Three things seem to be essential: First, the choice of a research area or a particular project. It must be scientifically ripe enough for applications and must also provide sufficient open questions of fundamental importance. Second, a particular structure should be created for such collaborations, one that will be undertaken for a specific time period and with specific goals and expectations. In this regard it is a good idea to operate as a group— or, as is frequent in the United States, a "center"—that can cut through the university's departmental structures and include all participants in a single unit. Other obligations within the university structure can remain unaffected by this. Such a special structure permits a common management of the group by industry and university together. Third, it is necessary that both parties understand what they hope to obtain from this collaboration and that they support each other in the mutual achievement of their goals. Only then can there be a harmonious climate that encourages the unexpected.[16]

## Overcoming the Innovation Deficit— With Help from the Universities

There are several quite concrete, and perhaps also very effective, opportunities for collaboration between pharmaceutical firms and university laboratories. We have already mentioned that as a whole, industry is not producing a sufficient quantity of substances for development, in large measure due to working on the wrong projects and spending too much time on them. Recall the simple equation

$$\frac{D \times \mu r}{t} = E,$$

where $D$ is the number of preclinical projects, $\mu r$ the probability of a project proceeding from research into development, and $t$ the average length of the research phase of the project. $E$ represents the number of resulting substances that go into development, $t$ is four years or longer, and $\mu r$ has historically hovered at around 0.4. If it were possible to make $t$ considerably shorter and to raise $\mu r$, then a given supply $D$ of preclinical projects would result in a higher number $E$ of development projects. If a firm were working on sixty projects, say, in its preclinical portfolio, then under the above assumptions they would be able to produce six development substances per year:

$$\frac{60 \times 0.4}{4} = 6.$$

If it were possible to shorten the research phase even more, e.g., to two years, and at the same time raise the probability of a substance going into development, say to sixty percent, then a much better result could be achieved, namely, the production of eighteen development substances:

$$\frac{60 \times 0.6}{2} = 18.$$

Yet how might it be possible to decrease the research time while increasing the probability of success? According to the above model, after two years all completed projects (fifteen per year) would have to be replaced with new "preincubated" ones. This could be achieved through extensive collaboration with university laboratories. If on average seventy-five projects were preincubated for two years and if the probability of such a project becoming a regular research project were approximately forty percent, then we would have our fifteen preincubated projects per year. Within four years, then, a firm's research productivity could rise from an annual six development projects to eighteen—theoretically in four steps: nine after the first year, twelve after two years, fifteen after three years, and, finally, eighteen after four years. Since financing laboratories outside of industry is considerably less expensive than performing work in-house,

such an increase in productivity could occur without a corresponding increase in costs. If one were to finance seventy-five such "external" projects
in steady state, then the total cost given an average of ten workers on each
project would be about $60 million annually. At present, the difficulty lies
in identifying suitable projects within the universities and in structuring
the flow of preincubated projects in such a way as to maximize each project's profitability. A further problem is the careful monitoring of these collaborative projects (Figure 7.2).[17]

Many American universities are aware that they are engaged in research activities that properly managed could be brought close to industrial development. At present there is talk of financing such activities

**Model for Collaboration Between Industry and Universities**

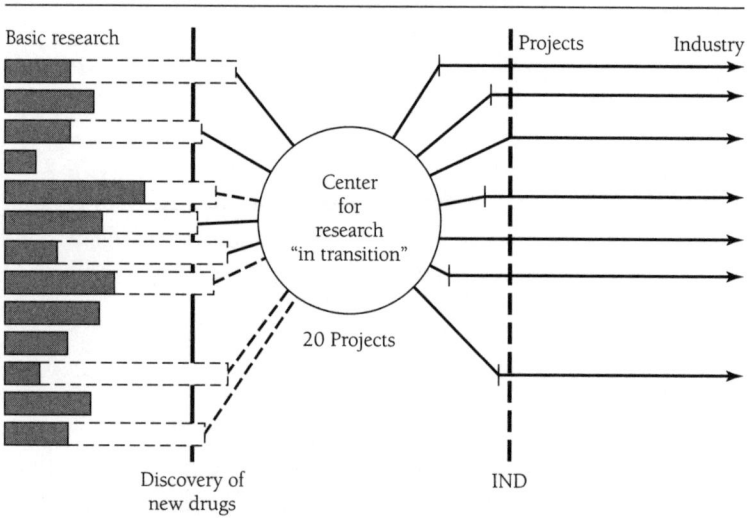

Figure 7.2.   The model is based on the following considerations: A firm
searches among a university's research projects for those that they believe may
lead, with additional work, to starting points for drug research. The basic-
research aspects of the individual projects is left untouched (crosshatched areas).
The company finances up to one hundred percent of the work that interests them
(area within dotted lines). The intellectual rights remain with the university,
while the company receives an exclusive license for the commercial use of all discoveries resulting from the collaboration. The university scientists retain complete use of the results for all internal, that is, scientific, purposes. A center for
"transitional research," essentially a small logistic tool, coordinates the contacts
between university and company research as well as the transition of projects
from the university into the company's research and development portfolio.

through special investment funds and then, when a project has progressed sufficiently, turning it over to industry at a profit. Such an arrangement would be attractive to industry, as it should to universities as well. But the latter must be careful not to lose sight of their primary mission: to produce knowledge and to disseminate it freely.

In the United States, universities are moving in the right direction to understand and master this difficult balancing act. Less positive is the situation in Europe, particularly with respect to the German university system. Many conditions that American universities have established for encouraging new industries and working flexibly and productively with existing firms are almost totally lacking in the German-speaking countries. For decades, the German university system remained tightly closed to the ideas of competition among universities, objective assessment of research achievements, and all measures directed at generating top quality in science. A more productive and competitive system with several independent professorships exists only in a few cases (the Heidelberg Center for Molecular Biology; the Gene Center of the Maximilian-University, in Munich). The old structure centered on a few professorial chairs with all its traditional dependencies is still the dominant arrangement. Continuing productivity and internationally recognized accomplishments play practically no role in maintaining an established professorial chair. Budgets are managed according to antiquated cameralistic principles of finance. Equipment cannot be exchanged for personnel and vice versa. Positions cannot be exchanged one for another except under the rarest of circumstances. Income, particularly in the clinical area, is unequally and unfairly allocated. We could cite many more such shortcomings. Taken together, they give the strong impression that the German university system as an institution is in desperate need of reform (one might also say in *chronic* need of reform). As it is currently constituted it can provide only limited encouragement for the establishment of new enterprises. It must improve its scientific basis as well as—and this above all—its institutional flexibility if it is to enter with industry into open and flexible partnerships that will contribute to national economic health through a rapid transfer of technology. In any case, we may hope that increasing pressure from Britain and the United States will compel the European universities to redefine their role in more imaginative ways than they have done up to now. Basic research and teaching are uncontested functions of the universities. But in addition, an active role in the transfer of basic knowledge into marketable products and processes must also become an essential concern of the universities. The best examples from outside Europe show that such a duality occasionally produces tensions; but they also demonstrate that the traditional role of the university and its recently expanding participation in technology transfer can be reconciled.

## Future Pharmaceutical Scenarios

We have diagnosed the current state of the pharmaceutical industry. The industry represents a unique worldwide collection of various capabilities and skills accumulated over more than a century, directed toward the goal of discovering and developing medicines. The large international pharmaceutical firms are still capable of balancing large research and development costs through timely international registrations in many countries and through a vigorous international marketing program. Smaller firms do not possess these options, and so they must seek partnerships. By and large, the industry finds itself under pressure from the demographic, economic, and scientific factors that we have already mentioned. If it is to survive, it must increase productivity decisively. Not all firms will be able to do so, and thus the number of profitable pharmaceutical firms will further decline. It is highly probable that some large firms will be able to further increase their achievements in development and production, and in the international marketing of their products. Whether they will also be able to do likewise with research is a matter of some doubt.

In recent decades, the intellectual—not necessarily the physical—conditions for research in the pharmaceutical industry have been worsening almost continuously. This statement may not apply equally to all enterprises, but on the whole it is correct. To the degree that industry gradually comes under public criticism and economic pressure, its relationship to its own research and development organizations will change. The blame for the discrepancy between what was available and what at the time might have been required to meet public criticism or economic pressure was often laid at the door of research. And this was not entirely unjustified. For decades many firms offered sinecures to scientists of only average creativity in which they could pursue their hobbies without having to subject themselves—unlike their colleagues in universities or publicly financed institutions—to international evaluation and criticism. A complacent mediocrity was the norm for many firms—so long as they were making a good profit. We should not fail to mention that this was also the period in which individual scientists like Gerhard Domagk, James Black,

Leon Sternberg and Lowell Randall, Paul Janssen, Gertrud Elion, George Hitchings, and Jean Borel achieved brilliant results. The creative achievements of these scientists were unplanned—could not have been planned. They did not arise from any entrepreneurial or strategic concepts and had nothing whatsoever to do with company politics.

## Researchers as Functionaries

Here we shall put forward the thesis that the type of creative individualist who may be entrepreneurial but who seldom thinks commercially, who is a priori motivated, who pursues a scientific problem because he sees within it a therapeutic possibility—that such a type is being driven out of today's pharmaceutical industry. In this person's place are to be found a steady supply of functionaries who can parrot the strategic maxims that they have been fed from one or another layer of management: individuals who are perhaps technically competent, but who fail to see the larger picture, who have no deep connection to scientific questions, but who opportunistically undertake whatever tasks are assigned to them. In short, the pharmaceutical industry is on its way to replacing its research organization with a technical apparatus capable of carrying out analyses, animal experiments, and syntheses, but incapable of developing new ideas or concepts. The research organizations of the large firms no longer direct their own operations. They are directed by lawyers, finance experts, salesmen, market strategists—individuals, that is, for whom the future is merely a linear extrapolation of current developments. We do not wish to give the impression that this development is irrevocable or irreversible. However, it must be pointed out that the pharmaceutical industry has created conditions that select against scientific individuality, creativity, and originality in favor of consensus, getting along, subordination, and a task-oriented mindset. There are still Domagks, Janssens, and Blacks to be found, but their successors are the founders and scientific directors of small, newly established enterprises whom we should seek primarily in the biotechnology industry.

## Reforms from Within and Scenarios for the Future

Can these trends be reversed? Can the industry use its newly won insight that research requires a different environment from that provided by today's pharmaceutical industry to organize its research program on a totally new basis and to liberate it from suffocating corporate structures?

This would require that from the top down pharmaceutical firms establish small research units with some of the characteristics of biotechnology companies. Will they be able to do this? Indeed, they could, if there were managers to be found who understood the problem clearly and who also had the authority to do something about it. With one or two exceptions it is doubtful whether such managers exist. And so matters will take their course. Actual research, that is, research directed toward the invention or discovery of new substances, will be found more and more in the small biotechnology firms, which, though they call what they do "biotechnology," are actually engaged in a broad spectrum of biological and chemical work. Of course, every year the prophets of the pharmaceutical industry renew their predictions of collapse for these younger siblings. Yet up to now this has not happened. This collapse will in fact not take place, because the biotechnology industry is increasingly taking up what the pharmaceutical industry is neglecting, namely, carrying out future-directed research, thereby creating the conditions for the discovery of new classes of medicines that attack more directly the root causes of disease. In many cases they are themselves making prototypes of such medicines available. A possible future scenario could well be one in which we have a consolidated, that is to say smaller, pharmaceutical industry engaged almost exclusively in international drug development, production, and marketing. A research industry arising from the biotechnology industry will supply these developers and distributors with ideas, concepts, and, above all, substances to develop (Fig. 7.3). A second, from today's point of view more optimistic, scenario might envisage that ten or twenty large international pharmaceutical firms will read the handwriting on the wall, abandon the path they have been following of overplanning and enveloping research in bureaucracy, and thereby recapture the lead in drug research. The biotechnology industry would remain a partner, continuing to specialize in particular technologies and leaving the actual integration of methods and products to the pharmaceutical firms.

A third scenario, somewhere between the previous two, sees a few firms remaining research-oriented and research-driven enterprises, with the remainder going the way of development and distribution. The biotechnology industry will develop into a "discovery industry," while the large firms continuing to do research will find new forms of collaboration with universities. Although maintaining a strong research profile has certain financial disadvantages (fixed costs) for a pharmaceutical company, there are some considerable advantages. If a firm is itself engaged in re-

**Future Scenarios for Pharmaceutical Research and Development**

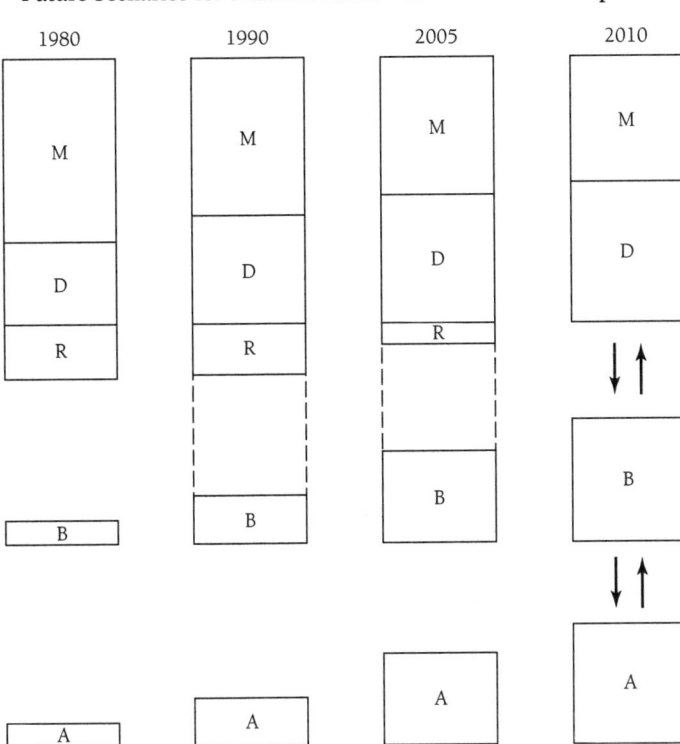

M = Marketing
D = Development
R = Research
B = Biotechnology industry
A = Academic centers

**Figure 7.3.** The model proceeds from the assumption that the research efforts of the pharmaceutical industry will gradually be transferred to the biotechnology industry, which itself is developing into a research industry. The significance of university research for drug research is growing.

search and has on staff specialists in particular important areas, above all scientists capable of integrative thinking, then it is better equipped than its non-research-oriented competitors to evaluate the worth and the risks of new technologies and new substances offered for development from outside sources. Furthermore, such companies will be the preferred partners of the discovery industry for collaborations in the difficult transition from research to development, ahead of those firms engaged only in development and distribution.[18]

## Scientific Possibilities: Economic Pressures

A further point of view should be mentioned. An industry under economic pressure concentrates more and more on therapeutic indications that promise high returns if a successful remedy is developed. Medical needs that cannot be translated into profit will no longer be worked on (tropical diseases, for example). Taking into account that in the industrialized countries an increasing portion of the population is uninsured (more than thirty million in the United States—about twelve percent of the population), there is the possibility that diseases affecting primarily the poor or less privileged layers of society will receive scant attention from the pharmaceutical industry. This means that the pharmaceutical industry will withdraw itself further from its self-appointed task of providing society with safe and efficacious drugs, perhaps further than society will tolerate. What might be the response to such a situation? Government or, better, university institutes for drug research that will have the skills to carry out all such functions? A new niche industry that can operate under different financial conditions from those of the pharmaceutical industry, which has become too large and too global? There are various possibilities. Certainly, drug research in the next century will no longer be the monopoly of the pharmaceutical industry, as it was in the latter half of the twentieth century. The therapeutic challenges are too diverse, and too diverse also are the economic contingencies of populations that must be provided with medicines. Finally, the scientific repertoire will expand considerably: gene therapy, modern vaccines, cell therapy, and xenogeneic transplantation will play a role and complement treatment with drugs. Of course, from these and other new technologies new needs for new drugs will arise. We have already indicated that research and development costs of several hundred million dollars per medicine have prevented firms from developing medicines that fail to promise annual returns in the hundreds of millions of dollars. If such a trend continues, will it not lead to an absurd situation: that the industry excludes itself from exploiting therapeutic possibilities? Some experts on health care believe that development costs could decrease significantly and that indeed this must occur if opportunities for the development of important new medicines are to be maintained. Scientific developments point clearly in the direction of a greater selectivity of medicines. Genome research above all will provide us with methods for producing remedies targeted at specific patient populations.

Hypertension and diabetes mellitus type II are probably not single diseases, but each of them is rather a disease phenotype that arises through a variety of genetic mechanisms. If every form of every disease could be paired with its specific treatment, then from the medical standpoint we should welcome this selectivity. But is it economically viable? Under today's conditions, decidedly not. A further example of the discrepancy between what might be desirable in the abstract and what is scientifically and economically achievable is that for decades the research drug firms have outdone themselves in the development of broad-spectrum antibiotics, cheered on by a host of clinicians. Today, it is not only the experts among microbiologists and clinicians who know that for broad-spectrum, and also broadly selecting, antibiotics there is a price to be paid, namely the existence of populations of multiresistant bacteria that in the larger clinics at least call into question the progress that has been made in the treatment of bacterial infections.[19] From the epidemiological as well as the clinical standpoint it would be preferable to have highly selective antibiotics against specific pathogens. These exist in large numbers, as every microbiologist knows. But as long as there is no diagnostic method that can *quickly* identify the one or more pathogens responsible for an infection, clinicians will prefer to administer antibiotics that act across a broad spectrum. Antibiotics effective across a narrow spectrum are hardly ever developed. But this could change under the influence of the polymerase chain reaction (PCR). This method makes it possible to copy specific sections of genetic information (RNA or DNA) in large quantity. In the realm of genetic information this method has the same significance as the photocopier had for printed information. In principle, it should be possible with the PCR method to identify all pathogens within a matter of hours and determine whether they are susceptible or resistant to the usual antibiotics. Narrow-spectrum antibiotics make sense in the context of PCR diagnostics, though it is doubtful whether the industry will make use of this connection. They are too deeply bound to their self-inflicted (or at least partially self-inflicted) economic constraints. It is furthermore unlikely that drug firms in the next century will play the same role as they did at the end of the twentieth century (Fig. 7.4). Drug research is about to become the concern of a variety of institutions, of which the pharmaceutical industry is but one. How central its role in the future will be depends on the flexibility and imagination that pharmaceutical managers bring to their industry and with which they react to societal changes and needs.

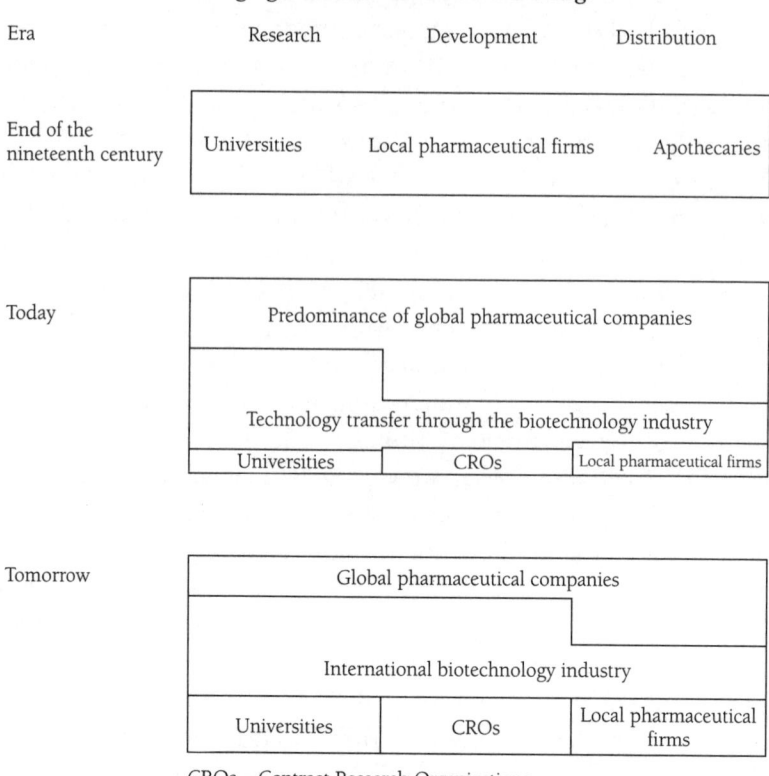

Figure 7.4.    During most of the twentieth century, the pharmaceutical industry possessed an almost complete monopoly on drug research and development, as well as on the manufacture of medicines. Today, it is in the process of losing this monopoly.

## Managed Care: The Union of Diagnosis and Treatment

In today's health-care systems one can see, particularly in the United States, the desire on the part of hospitals and other health-care providers to offer a variety of mutually agreed-upon services from a single source, and such arrangements are now coming into existence. For instance, important medicines against HIV infection as well as the widely accepted test based on the polymerase chain reaction, which provides a quantitative measure of the viral load and thus allows for a continuous control of therapeutic success, come from two divisions of the same firm. The possibility of providing a complete package of related measures by a single firm represents a further challenge that the pharmaceutical industry will face in the coming century.

## Connections Between Basic Research and Pharmaceutical Innovation

From an anthropological point of view, the practice of science comes out of two possible impulses. One is a desire to *understand* the world and its manifestations, the other to make the world more habitable and controllable for humankind, that is, to *change* the world. Both motives have been active throughout history. At first, over a long period of time, each developed its own independent traditions. In ancient times the world of science belonged to the philosophers and mathematicians, while later in Western culture to the Scholastic philosophers and theologians, then to the Humanists, none of whom had an interest in employing scientific methods to alter their environment. They represented the spirit that founded the universities, which even today have as their primary goal the production of new knowledge and its dissemination without regard to practical applications.[20]

In contrast, there was a parallel tradition of inventors and craftsmen, that is, of architects and builders, draftsmen, weapons builders, engineers in the broadest sense, whose work was based more on empiricism than on theory. It was only in the nineteenth century that natural science had developed to where it could serve as the basis for further development of technology. That is, technology was placed on a firm scientific footing.[21] The commanding position held by technology at the close of the twentieth century is a direct consequence of this development: Railways, airplanes, medicines, computers, modern techniques of construction and lighting, all are achievements based on chemistry, physics, and biology. Today, the development of technology is closely bound to scientific progress; indeed, they can no longer be separated. Thus in the twentieth century a consequence of technology's being placed on a more scientific basis has been that science has become more technological. That is, progress in science has become dependent on the development of new technological tools. For example, modern biological research is unthinkable without spectroscopy, ultracentrifugation, electrophoresis, electron microscopy, and many other aids to research that came originally out of physics or chemistry. Thus today we are in a situation where the categories of thought and action, represented by science and technology, which developed apart before the nineteenth century, are bound together in a variety of ways. On the one hand, the theoretical maturity of basic research made possible applied research—that is, research focused on goals external to science—while on the other hand, basic research is more and more dependent on the results of applied research and the contributions of scientifically based technology. Understanding the universe and trying to change it—basic research and applications—are thus no longer separate realms: They are mutually dependent.

Yet the original motives are still at work. The world becomes accessible in a different way to a type of research that concerns itself primarily or exclusively with understanding the ways in which things are connected, rather than to a type of research whose goals lie outside of science. Although science and technology are close to each other, they proceed as always from different premises. Such a distinction holds analogously for basic research and applied research.

## Thought and Action in Drug Research

By its very nature, pharmaceutical research falls into the category of applied research. Its goal, the cure or amelioration of disease, is an extrascientific one. At its core it has little to do with a desire to understand the natural world. In practice, however, drug research relies on the findings of basic research. Conversely, the discoveries of applied research provide tools for explaining basic biological processes. The antibiotics are an example. They were first described as the material cause of a natural phenomenon, which in 1878 Paul Vuillemin called "antibiosis."[22] In the twentieth century, many "antibiotic" substances were sought, found, and developed into medicines. The study of their modes of action provided much information. The great majority of antibiotics inhibited key biosynthetic processes in microorganisms. As specific inhibitors of protein or nucleic acid synthesis these materials became important tools in the hands of biochemists. Without them, fundamental cellular functions could not have been explained, unless perhaps by a very circuitous route. Thus basic research led to the establishment of a principle that was then applied to a medical problem, namely the treatment of infections. The drugs thus found then turned out to be fundamental tools of basic research. This cycle of discovery–application–renewed discovery occurs today with much greater vigor and much more rapidly, as can be observed in the example of antibiosis and antibiotics.[23] Antibiosis was first described at the end of the nineteenth century (Pasteur and Joubert, 1877). But it was only in 1939 that Ernest Chain and Howard Florey undertook a serious attempt to find an application of this principle to medical practice. That in the process they resorted to penicillin, discovered by Fleming in 1929, was a lucky circumstance. The span of time between important breakthroughs in basic science and the corresponding practical consequences for diagnosis and treatment has become shorter today than what we saw in the example of antibiotics: In 1973 Herbert Boyer and Stanley Cohen reported on the first experiment in which a mammalian gene was cloned and expressed in bacteria. Only twelve years passed before the registration of recombinant human insulin. In the case of monoclonal antibodies, which were discovered in 1975, the time was even shorter. After the Ob gene was cloned, sequenced, and expressed in a variety of cells in 1995, it was just a matter of weeks before research teams in industry and universities tested the therapeutic suitability of leptin, the protein coded by the Ob gene. Leptin turned out to be a hormone that governs food intake and so may well play an important role in the treatment of obesity. The dialogue between basic research and applications has become quite intimate. Applied research, and thus drug research as well, cannot exist without constant dialogue with basic research. The functional proximity of the two types of research demands a greater institutional and, ideally, spatial proximity. In many firms it is not understood that research that concentrates exclusively on goals outside the purely scientific—in this case therapeutic goals—is in danger of becoming ossified and formulaic, losing all originality. And indeed, we can trace the low degree of scientific productivity, a problem that plagues the pharmaceutical industry, to its uncertain and insufficiently nurtured relationship to basic research. Whoever believes that basic research in relation to drug research is a luxury has not understood that research that concerns itself exclusively with goals that lie beyond the scientific horizons will sooner or later dry up.

## How Can Applied Research Renew Itself?

What should be the relationship between pure and applied research? How can we ensure that pharmaceutical research will continually be able to recharge its batteries? The best way is to see to it that the groups responsible for particular drug-development projects are also carrying out basic research "nearby." Of course, the principal goal must remain the discovery and development of drugs. But a scientist working on a new molecule and its characterization should have the opportunity to pursue questions that go beyond this immediate goal. Ten to twenty percent of a scientist's time can and should be devoted to working on such basic questions. Often, such research leads to surprising insights, which in turn become new starting points for application-oriented work. Many firms, above all in biotechnology, offer their scientists the opportunity of keeping ten to twenty percent of their time free, that is, unconnected to any project and to be used at their own discretion.[24] Other firms have no official rules, but they encourage scientific work that lies outside the immediate goals and requirements of the project. But in general, the relationship between drug research and biological basic research is ignored. Of course, such excursions into biological or chemical research that occupy only twenty percent of an individual's workload are insufficient for an individual scientist to do first-rate work in these areas. But they are certainly sufficient for undertaking collaborative work. Through alliances with basic research in universities as well as in public or private institutes, researchers in industry can certainly obtain access to the newest ideas and technologies and to new developments in basic science.

## Industry-Related Basic Research

Of course, there are additional opportunities to unite basic research and drug research effectively. One possibility consists in the pharmaceutical industry financing departments or institutes that carry out basic research in spatial and thematic proximity to their drug research. The Friedrich Miescher Institute, in Basel, which was founded by Ciba; the Basel Institute for Immunology; the former Roche Institute of Molecular Biology, both established by Roche; or DNAX, in Palo Alto, an institution of Schering Plough—these are all examples of such a strategy. The problem is not so much the establishment and maintenance of good scientific institutes. These are, of course, not trivial tasks. However, it is more difficult to associate such an institute with the parent firm's research in such a way that maximal use of it is made by the financing firm but the scientific independence of the institute is not jeopardized. Only this independence can enable an institute to attract first-rate scientists in a specific field. Of course, this same independence that attracts first-rate researchers harbors the danger that they will hold themselves aloof, thematically and operationally, from their colleagues in drug research. This dilemma can be resolved only by *informal* methods. The secret lies first of all in the choice of subject area for such an institute, and also in the choice of a director or directors who will bear responsibility for the institute's fate. An area must be chosen that is of wide-ranging strategic interest for the firm. For example, immunology was such a theme for Roche in 1969. The area must be mature enough that central questions seem to be answerable in the near or at least not-too-distant future (within a decade), and it must provide concrete points of departure for drug research. Equally as important as the choice of area is the choice of directors of such an institute. One should select only scientists who alongside their passion for basic research also possess a genuine interest in applications and who enjoy working with biologists and chemists engaged in applied research. Closeness cannot be forced from without. It must arise of itself internally based on the chosen area of research and the organization of the individuals involved. If everything is in order in this regard, then the institutes or departments for basic research connected to a pharmaceutical firm can represent an enormous gain. But should a single critical element be lacking, the whole enterprise can lead to disappointment. Institutes established for carrying out basic research are expensive. They are worth the expense only when the basic requirements—the research area and the directors of the institute—have been well chosen.

Other opportunities for combining more basic research with drug research are to be found in collaborations with universities and other institutes devoted to basic research, though there is always the danger of slipping into old patterns, repeating the conventional roles of university and industry. The same rules hold for collaborations with universities in strategically important areas of research as for collaboration in more applied areas: One must define what both parties expect from the collaboration, and at least for more extensive collaborations one should establish a joint management structure consisting of a researcher from industry and a colleague from the university.

## Sabbaticals

Finally, in this connection we should make reference to the academic custom of the sabbatical year. Professors in British and American universities generally have the opportunity every six or seven years to spend a year in association with another university or research institution, either in their own country or abroad. In industry such a creative break, which gives the scientist the opportunity to recharge his batteries and obtain renewed contact with developments in his field of research or a neighboring field, is especially important. For those who work in interdisciplinary configurations on research whose goals are not purely scientific, it is necessary from time to time to pause so that new approaches can be found and ideas developed. If the concept of a sabbatical year for researchers in industry has not found the favor it deserves—at least in the author's experience—there is an important reason: A prolonged absence of six months or a year in the competitive atmosphere of industrial research almost always necessitates that another colleague take over the work of the absent researcher. Upon returning from a sabbatical, the researcher often finds the position and functions that he left behind taken over by a colleague, and he must—often with great effort—forge a new place for himself in his old organization. Flexibility is not the pharmaceutical industry's strong suit, flexibility with personnel especially, and the industry's hierarchical structures make reintegration especially difficult. In smaller research units such models have sometimes functioned well, indeed made it easier to change fields and begin new projects for which important theoretical principles need to be "imported" from without. The author recalls in particular a case from the mid-seventies when a pharmacological research center had to find a new focus for its research. Research in parasitology was abandoned, as well as in animal nutrition and veterinary medicine. In their place came the new fields of antifungal chemotherapy, immunopharmacology, and a modern bacteriology that also extended to chlamydiae and other pathogens of sexually transmitted diseases. Without the extensive use of well-planned research residencies in the world's best laboratories—sabbaticals, in short—the reorientation, in which no one was laid off, would not have been possible. It was completed in a little over a year, and it proved very successful, especially in the fields of mycology and immunopharmacology. This tool of "personal and scientific renewal" should not be forgotten today.

## Gambling Away the Future

The future of pharmacological research will be shaped by many forces, which we can place in two groups. The first group we might call the scientific–technological cluster: To this group belongs the unpredictable course of science itself; secondly, it includes the question of the effectiveness of technology transfer from the universities to industry. Thirdly, we should mention the universities' conception of their mission, which in the German-speaking countries, at least, requires a new formulation. Fourthly, we could cite the renewal of applied research by the results of basic research, which must take place constantly and which ideally has in turn a positive effect on basic research.

A second cluster of influences can be collected under the rubric of scientific and social causes. Here belong macroeconomic developments and the future direction of health-care markets. Will the American principle of managed care gain ascendancy? Will the consumer, ultimately the patient, expect to receive the provision of a variety of health-oriented services from a single source? Or will therapeutic and diagnostic specialists, those with specialized knowledge of equipment, transplantable cells, tissues, and organs, assert themselves in the marketplace? There will certainly be firms for xenogeneic transplantation, firms that offer skin and tissue transplantations, as well as those offering marrow transplantations. Such products require completely different methods of production, other forms of quality control, and different marketing and sales strategies from those known in today's pharmaceutical and classical biotechnology industries. Of course, it is possible that large pharmaceutical firms will acquire these newly evolving areas through direct purchase. As of today, we know too little about the institutional embodiment of these new medical measures to speculate about it with any degree of confidence.

A multitude of factors will affect pharmaceutical research and the pharmaceutical industry, and it is difficult, if not impossible, to make accurate predictions about the fate of drug research and the future of the pharmaceutical industry. The interaction of so many scientific, economic, and sociopolitical factors is reminiscent of the moves in a complex game. Probably, many models concerning industry's role in providing treatment, diagnosis, and perhaps also prevention will be tried over the next several years. Some will turn out to be worthwhile and form the basis of new industrial configurations. Others—the majority—will be rejected. Like biological evolution, the evolution of new industrial forms is a game in which new forms arise by chance variations and selection, and then become obsolete when conditions change. This is the nature of human institutions, and so it will continue to be in the future.

Predictions are possible only over a short time span, say, over ten, or at most twenty, years. These numbers also represent the time it takes for a new technology or a new scientific concept to be transformed into marketable processes or products. What will be reality in ten or fifteen years must be recognizable today at least in its initial stages, not necessarily, of course, to everybody, but to those who have developed a broad and sensitive understanding of science and technology and the possibilities for their application. If the author assumes himself to possess such attributes—and he could, of course, be mistaken—then two things are obvious: Firstly, the complexity of the technological and scientific activities that are necessary for the discovery and development of a new medicine are such that in the near future there will remain very few institutions capable of uniting all the necessary capabilities under one roof. Furthermore, specialized industries will develop that will play the role of suppliers to the large pharmaceutical firms. They will supply ideas, technologies, and new products. Secondly, almost all of the large pharmaceutical firms tend toward centralization, toward the integration of research and development into the unified structure and function of a single firm or division. They seem—perhaps under the inevitable group dynamics of large organizations—to wish to continue along this path. Such a path, however, represents the subordination of research to particular strategies determined by market forces and the enterprise as a whole. Even if some of the more sensitive managers of large pharmaceutical firms are aware of the relationship of research to the rest of the firm, the majority are completely in the dark, and therefore the trend that is underway will continue. But this can lead to a situation where the large pharmaceutical firms lose a considerable portion of their pharmaceutical research; they will be simply throwing it away. What, then, will it mean to subordinate research to the requirements of the market? It means that long-range problems have been subordinated to short-term needs, that the innovative must take a back seat to the tried and true; but above all, this subordination of research signals a reversal of the classical roles played by research and marketing. The contributor of ideas—research—will become a receiver of instructions, while the implementer of new concepts in the marketplace—marketing—will become the contributor of ideas. But these ideas won't go very far, since they reflect the marketplace, which is to say that they reflect today, if not yesterday, and never the truly novel. This reversal of competencies is incompatible with original research. Although technical competence, knowledge of the literature, and technological skill might continue to exist even under these reversed conditions, the creation of the truly original will not. In this sense the notion of "game" in the sense of a future determined by chance has another meaning: As in a game of chance, the classical pharmaceutical industry is in the process of gambling away its future as a *research* industry. Research, and drug research in particular, will, of course, continue. But in the twenty-first century it will no longer be a monopoly of a single industry, certainly not the monopoly of the classical pharmaceutical industry. Many partners will play a variety of roles in this new game: the biotechnology industry as the actual discoverer, the universities as important providers of ideas and preincubated projects, the classical pharmaceutical firms as developers, manufacturers, and distributors, and, increasingly, contract research organizations (CROs) as supporters of development.

"*Les jeux sont faits!*" cry the croupiers as the roulette wheel turns. For the pharmaceutical industry many wheels are turning. It is a complex game. When the wheels stop and the balls have dropped into the winning numbers, there will be several winners. Drug research, however, especially its creative and original aspects, will have to find itself a new home. To some extent, it would seem, it already found one.

# ⁓⁓⁓ Chapter Notes

## Chapter 1

1. Schumpeter, J.: *Business Cycles: A Theoretical, Historical and Statistical Analysis of the Capitalist Process*. McGraw-Hill, New York 1939.

2. Horisberger, B.: Kosten-Nutzenanalyse der Ulkustherapie. *Schweizerische Medizinische Wochenschrift* **114**, 699–706 (1984). See also *Grundlagen der Arzneimitteltherapie*. Dölle, W., Müller-Oerlinghausen, B., and Schwabe, U., eds. Bibliographisches Institut, Mannheim, Wien, Zürich, 1985.

3. Drews, J.: The influence of drug research on concepts of disease in clinical medicine. *Drugs made in Germany* **30**, 47–52 (1987).

4. Bonnameaux, H., de Moerloose, and A. Campana: Oral contraception and menopausal hormone replacements: Effects on hemostasis and risk of venous thromboembolism. *Schweiz. Med. Wochenschrift* **126**, 1756–1763 (1996). Kanis, J.A.: Estrogens, the menopause and osteoporosis. *Bone* (Suppl. 5) 1853–1905 (1995).

5. Lukas, S.E.: CNS liability of anabolic–androgenic steroids. *Annual Review Pharmacology* **36**, 333–357 (1996). Lloyd, F.H., Powell, P., and Murdoch, A.P.: Anabolic steroid abuse by body builders and male subfertility. *British Medicine Journal* **313**, 100–101 (1996).

6. Rohypnol and Rape. Internet: Alta vista web pages on "Rohypnol and rape." http.//www.ocs.mg.edu.an/korman/feminism/rohypnol/html.

7. Drews, J.: Orphan drugs—the European perspective. *Drugs made in Germany* **31**, 8–9 (1988).

8. A European "Orphan Drug" law is now (as of March 1997) being prepared by the European Commission. The present draft contains the same precautions and incentives as the American law and indeed anticipates an even longer period of exclusivity than the seven years of the American statute.

9. See also Drews, J.: *Arzneimittelforschung und ethische Verpflichtung*. Editiones Roche, Basel 1994, pp. 16–17. Serious programs for discovering and developing

effective remedies against tropical diseases are no longer part of the agenda of the pharmaceutical industry. Nonetheless, there have recently begun discussions aimed at cooperation between the pharmaceutical industry and international granting organizations such as WHO and the World Bank toward the discovery and development of new drugs against malaria, leishmaniasis, trypanosomiasis, and other diseases. One concrete suggestion in this direction is discussed in the publication cited.

10. The American firm Bristol-Myers was long considered such a "specialist" in this area, achieving a very strong position through its own research, but even more so through the systematic licensing of new substances. Modern techniques for the preparation of recombinant proteins and monoclonal antibodies have leveled the playing field. $\alpha$-interferon, and antibodies against B-cell lymphoma and against growth-factor receptors, have been and continue to be developed by biotechnology firms. These developments have provided a firmer basis for the medical treatment of cancer.

11. At the time of writing (1997) there are three protease inhibitors on the market. They achieved the following sales figures in the year 1996 in the USA alone: Sequinavir: $147 million (introduced December 1995); Indinavir: $120 million (introduced March 1996); Ritonavir: $46 million (introduced March 1996).

12. Contract research organizations (CROs) are firms that offer their services in carrying out particular segments of the drug-development process. At first the pharmaceutical firms considered these newcomers to be insufficiently qualified. However, these companies have developed into highly qualified "specialists." Many of them concentrate on clinical development (Parexel, Besselaer), while others also offer preclinical development (Quintiles). The CROs have become important partners for today's pharmaceutical industry. They have allowed the pharmaceutical firms to reduce their staffs by outsourcing a considerable portion of their own development work. This works as a form of insurance against excess capacity in cases where a research project collapses.

# Chapter 2

1. See in this regard Ritter, P.: *Im Spiegel der Arznei.* S. Hirzel, Stuttgart 1990, p. 32. In Prussia, between the years 1822 and 1855, the number of apothecaries was less than one per ten thousand inhabitants. In the western part of Germany the figure was somewhat higher (one apothecary per seven thousand inhabitants). Historians who study Prussian medical statistics have established that the licensing of new apothecaries did not keep pace with population growth between 1859 and 1867. This slowing down in establishing new apothecaries, for which the Ministry of Education is to be held responsible, was on the one hand the cause of hundreds of unnecessary deaths among the population, while on the other hand it encouraged the pharmaceutical industry to institute mass production and establish their own sales networks.

2. The development of apothecaries, the silk industry, and the aniline dye industry into pharmaceutical firms is described in the case of the Basel firms in

the following books and articles. Riedl, R.: "A brief history of the pharmaceutical industry in Basel" in *Pill Peddlers*, J. Liebenau, New York 1990; and Peyer, H.C.: *Roche—Geschichte einer Unternehmung 1896–1996*. Editiones Roche, Basel 1996.

3. Mahoney, T.: *The Merchants of Life: An Account of the American Pharmaceutical Industry*. Harper, New York, 1959.

4. Liebenau, J. et al., *Pill Peddlers: Essays on the History of the American Pharmaceutical Industry*. American Institute of the History of Pharmacy, 1990.

5. In many universities in Germany, Austria, and Switzerland there still exist traditional faculties in pharmaceutical science, with institutes of pharmaceutical chemistry, pharmacology, formulation science, and pharmacognosy. In 1996 a commission of experts suggested to the University of Heidelberg that its present department of pharmaceutical sciences be transformed into a modern center for drug research, in which all functions necessary for the discovery of new drugs would be combined and which would include new professorships and research groups. The establishment of such university centers could give drug research an important boost. In Chapter 7 we shall discuss the possibility that the pharmaceutical industry will lose its "monopoly" on drug research and that other institutions, including universities, will share in it. The construction of drug research institutes in universities would have great significance in this connection.

6. Ebbel, B.: *The Ebers Papyrus, the Greatest Egyptian Medical Document*. Leon and Muniksgaard, Kopenhagen, 1937. Leake, C.D.: *The Old Egyptian Medical Papyri*. University of Kansas Press, 1952.

7. An extensive presentation of the history of opium with references to the original literature can be found in Sneader, Walter: *The Evolution of Modern Medicines*. Wiley and Sons, Chichester and New York, 1985. See also Wright, A.D.: "The History of Opium." *Med. Hist.* **18**, 62–70 (1968) as well as the original paper of Sertürner, F.W.: "Über das Morphium, eine neue salzfähige Grundlage und die Mekonsäure als Hauptbestandtheile des Opiums." *Gilbert's Annalen der Physik* **25**, 56–89 (1817), Leipzig.

8. Pelletier, P.I. and Caventou, J.B.: *Ann. Chim. Phys* (Paris) **15**, 289–318, 337–365 (1820).

9. Walter Sneader: *Drug Discovery*. John Wiley and Sons. Chichester, New York 1985, pp. 13–14.

10. Charles Mann and Mark Plummer: *Aspirin*. Droemer/Knauer, Munich 1993. p. 34.

11. Wenckebach, K.F. Cited in Szekeres, L. and Papp, J.G.: "The Discovery of Antiarrhythmics," in *Discoveries in Pharmacology*. M.J. Parnham and J. Bruinvels, eds. vol. 2, p. 198. Elsevier 1984.

12. Theophrastus Paracelsus, *Werke*, Bd. I: Med. Schriften (W.E. Peuckert, ed.). Wissenschaftliche Buchgesellschaft, Darmstadt 1965.

13. These details were taken from G. Stille: *Der Weg der Arznei*. Braun, Karlsruhe 1994, pp. 231–234.

14. *An Account of the Foxglove and Some of Its Medicinal Uses*. G.G.J. and J. Robinson. Birmingham 1785.

15. Hufeland, W.: *Makrobiotik oder die Kunst, das Leben zu verlängern.* Wien 1832.

16. See Walter Sneader: *Drug Discovery.* John Wiley and Sons. Chichester, New York 1985, pp. 137–138.

17. Schönlein, J.L.: *Allgemeine und spezielle Pathologie und Therapie. Nach den Vorlesungen niedergeschrieben und herausgegeben. Litteratur Comptoir,* St. Gallen 1841.

18. Traube, L.: "Über die Wirkung der Digitalis, insbesondere über den Einfluß derselben auf die Körpertemperatur in fieberhaften Krankheiten." *Charité Ann. I. Ges. Beiträge zur Pathologie und Physiologie,* Berlin 1871, vol. II, pp. 97 *ff.*

19. Böhm, R.: "Untersuchungen über die physiologische Wirkung des Digitalis und des Digitalin." *Pflügers Archiv* 5, pp. 153–191 (1872).

20. Dreser, H.: "Über Herzarbeit und Herzgifte." *Arch. exp. Pathol. u. Pharmakol.* 24, pp. 221–240 (1887).

21. This information taken from W. Sneader: *Drug Discovery,* p. 138.

22. This description follows in broad outline that of Ch. Mann and M. Plummer: *Aspirin.* pp. 9–70 as well as Mann, J.: *Murder, Magic and Medicine.* Oxford Univ. Press 1992. See also Dreser, H.: "Pharmakologisches über Aspirin (Acetylsalicylsäure)." *Pflügers Arch. Anatomie u. Physiologie* 76, 306–318 (1899).

23. Ergot is the common name of *Claviceps purpurea,* derived from its shape, like a cock's spur (old French *argot*).

24. Moir, J.C.: "Ergot: from St. Anthony's fire to the isolation of its active principle, ergometrine (ergonovine)." *Am. J. of Obstetr. Gynecol.* 120, 291–296 (1974). A quite extensive presentation can be found in *Discoveries in Pharmacology,* vol. 2, in the article by Barbara Clark: "The versatile ergot of rye." pp. 3–36 (1984).

25. Huisgen, R.: "Richard Willstätter." *J. Chem. Education* 38, pp. 10–15 (1961).

26. Hofmann, A.: "The discovery of LSD and subsequent investigations on naturally occurring hallucinogens." In: *Discoveries in Biological Psychiatry.* Lippincott, Philadelphia 1970, pp. 91–106.

## Chapter 3

1. See Stille, G.: *Der Weg der Arznei.* Braun, Karlsruhe 1994, pp. 183 *ff.*

2. An extensive discussion of this connection is given by G. Cauguilhem in George Cauguilhem: *Wissenschaftsgeschichte und Epistemologie. Ges. Aufsätze.* W. Lepenius, ed. Suhrkamp Verlag, Frankfurt 1979, pp. 75–88. Cauguilheim points out particularly that unlike his predecessors and contemporaries, Bernard interpreted deviations and inconsistencies in the results of an experiment as evidence of unconsidered hypotheses, which he insisted must be looked for.

3. Schmiedeberg, O.: "Rudolf Buchheim, sein Leben und seine Bedeutung für die Begründung der wissenschaftlichen Arzneimittellehre und Pharmakologie." *Arch. expt. Path. Pharmakol.* 67, 1–17 (1911/12). Buchheim, R.: *Lehrbuch der Arzneimittellehre.* Leopold Voss, Leipzig 1878.

4. G. Stille: *Der Weg der Arznei.* pp. 208–209.

5. Koch-Weser, J. and Schechter, P.J.: "Schmiedeberg in Strassburg 1872–1918: The making of modern Pharmacology." *Life Sciences* **22**, pp. 1361–1372 (1978).

6. See Stille, G.: *Der Weg der Arznei.* pp. 223–224.

7. "Organic chemistry since 1860." In: *A history of chemistry.* F.J. Moore, New York 1918, pp. 212 *ff.* and "Kekulé: Molecular architecture from dreams," in Roberts, R.M.: *Serendipity.* John Witney and Sons, 1989.

8. "Natural and artificial organic substances." In: *The Evolution of Chemistry.* E. Farber, Philadelphia 1952.

9. See Farber, E.: *The Evolution of Chemistry.* pp. 178 *ff.*; see also "Synthetic dyes and pigments." In: Roberts, R.M.: *Serendipity.* pp. 66 *ff.*

10. Fracastoro, G.: *De Contagione et contagiosis morbis et eorum curatione, libri III.* translation and notes by Wilmer Cave Wright, New York, Putnam, 1930.

11. Popper, K.: *The Logic of Scientific Discovery.* New York, Basic Books, 1959 (1931).

12. All details given here from the history of chemotherapy are to be found in the papers of Paul Ehrlich, edited by F. Himmelweit. Ehrlich, P.: *Gesammelte Arbeiten.* F. Himmelweit, ed. Springer, Berlin, Göttingen, Heidelberg 1957. A more complete history than in the present book can be found in Drews, J.: *Grundlagen der Chemotherapie.* Springer, Vienna, New York 1979.

13. Drews, J.: *Grundlagen der Chemotherapie.* Springer, Vienna, New York 1979. In this book the relevant primary sources are cited.

# Chapter 4

1. The expression "antibiosis" was coined in 1877 by P. Vuillemin. He used this term to express the opposite of "symbiosis." A contribution to the intellectual history of antibiotic research from 1850 to 1950 can be found in the dissertation of Andrea Cardon de Lichtbuer: *Zur Entwicklungsgeschichte der antimikrobiellen Wirkstoffe.* Basel 1990.

2. Gibbons, A.: "Exploring new strategies to fight drug-resistant microbes." *Science* **257**, 1036–1038 (1992).

3. Monaghan, R. and Tkacz, J.S.: "Bioactive microbial products. Focus upon mechanisms of action." *Ann. Rev. Microbiol.* **45**, 271–301 (1990).

4. Meldrum, N.U. and Roughton, F.J.: "Carbonic anhydrase, its preparation and properties." *J. Physiol.* **80**, 113–142 (1933).

5. Maxwell, R.A. and Eckhardt, S.B.: *Drug Discovery: A Casebook and Analysis.* Humana Press, Clifton 1990.

6. Anthony Fauci: "AIDS in 1996. Much accomplished, much to do." Editorial. *JAMA* **276**, 155–156 (1996).

7. Steinmetz, M.: unpublished data. The analysis is from 1996. The separate categories should remain stable with respect to one another over the next three to five years.

8. Langley, J.N.: "On the reaction of cells and certain nerve endings to certain poisons, chiefly as regards the reaction of striated muscle to nicotine and to curare." *J. Physiol.* (London) **33**, 374–413 (1905).

9. Ahlquist: "A study of the adrenotropic receptors." *Am. J. Physiol.* I, 100–106 (1948).

10. Biotechnology Medicines 1995 Survey, Pharmaceutical Research and Manufacturers of America (PhRMA). Riethmüller, G., Schneider-Gaedicke, E., and Johnson, J.P.: "Monoclonal antibodies in cancer therapy." *Curr. Opin. Immunol.* **5**, 732–739 (1993).

11. Haseltine, W.: personal communication.

12. Drews, J.: "Genomic Sciences in the Medicine of Tomorrow." *Nature Biotechnology* **14**, 1516–1517 (1997). See also Drews. J. and St. Ryser: unpublished results, 1997.

13. Kuhn, Th.: *The Structure of Scientific Revolution*, second ed. University Press, Chicago 1970.

14. See also Drews, J.: *Naturwissenschaftliche Paradigmen in der Medizin*. Editiones Roche, 1992.

15. The observation and classification of changes in organs as the cause of disease had already begun in ancient Greek medicine. Erasistratos of Kea described cirrhosis of the liver as a morphological change and also recognized the connection between this deviation and the onset of an abnormal accumulation of fluid in the abdominal cavity (ascites). Galen of Pergamum united this tradition of organ and tissue pathology with that of the humoral pathology of Hippocrates and Aristotle. Although Galen always held to the Hippocratic concept of "hot, cold, wet, and dry," he knew that every disturbance of a bodily function must have its morphologically recognizable cause. See Kudlin, F.: *Griechische Medizin*. Zürich 1967; Galen: *Selected Works*, Oxford, 1997; and the Loeb edition of Hippocrates (4 volumes), translated and edited by W.H.S. Jones and E.T. Withington, Cambridge, Mass. Harvard Univ. Press 1948–1953. The attempt at a systematic representation of morphological causes of disease was, to be sure, undertaken much later. In his book *De abditis morborum causis* (Ch. Singer ed., Springfield, Ohio 1954) Antonio Benevienis (1448–1502) employed comparisons between clinical syndromes and morphological findings. Andreas Vesalius (1514–1564) set this hypothesis on a firm methodological footing. Vesalius performed his own dissections and studied changes in organs directly in the cadaver. He is considered the father of anatomy. But even Vesalius cannot yet be considered the founder of a comprehensive morphologically anchored theory of disease. He was still too tightly bound to Galen's humoral–pathological conceptions. Moreover, many of his anatomical findings had to do with *normal* anatomy and not with pathological anatomical changes. And finally, developmental biology and aspects of cellular biology were not yet present in his work. These components of a morphological theory of disease were not introduced until the eighteenth century. They are connected with the names J.L. Prévost, in France, and Karl Ernst von Baer, in Germany. With the development of stronger microscopes and new staining techniques histology developed very rapidly. Wilhelm von Waldeyer, the

teacher of Paul Ehrlich, showed, for example, that cancer cells have their origin in epithelium and not in connective tissue. All of these developments united finally in the cellular pathology of R. Virchow, whose maxim "Omnis cellula e cellula" became the basis of a morphological paradigm in medicine. One might formulate the theses of such a paradigm as follows:

1. Life is bound to structures, that is, to cells, organs, and organisms.
2. Structures develop in strictly regulated ways.
3. There are demonstrable relationships between structure and function.
4. Diseases can be recognized and interpreted as structural deviations from the norm.
5. Relationships between the structure and function of organs are therapeutically useful (for example, in surgery).

See also Drews, J.: *Wissenschaftliche Paradigmen in der Medizin*. Editiones Roche, Basel 1992.

16. Drews, J.: *Wissenschaftliche Paradigmen in der Medizin*. Editiones Roche, Basel 1992.

17. Avery, O., McLeod, C.M., and McCarty, M.: "Studies on the chemical nature of the substance inducing transformation of pneumococcal types." *J. Exp. Med.* **79**, 137–158 (1944).

18. Watson, J.D. and Crick, F.C.H.: "A structure for deoxyribonucleic acid." *Nature* **171**, 737–738 (1953).

19. Watson, J.D. and Crick, F.C.H.: "General implications of the structure of deoxyribonucleic acid." *Nature* **171**, 964–967 (1953).

20. Kornberg, A.: "DNA replication." *J. Biol. Chem.* **223**, 1–4 (1980).

21. Nathans, D. and Smith, H.O.: "Restriction endonucleases in the analysis and restructuring of DNA molecules." *Ann. Rev. Biochem.* **44**, 273–293 (1975).

22. Nirenberg, M. and Leder, P.: "RNA codewords and protein synthesis." *Science* **145**, 1399–1407 (1964); and Woese, C.B.: *The Genetic Code.* Harper and Row, 1967.

23. Sambrook, J., Fritsch, E.F., and Maniatis, T.: "Molecular Cloning." Cold Spring Harbor Laboratory Meeting 1989.

24. Sanger, F.: "Determination of nucleotide sequence in DNA." *Science* **214**, 1205–1210 (1981).

25. Maxam, A.M. and Gilbert, W.: "A new method for sequencing DNA." *Proc. Natl. Acad Sci. USA*: **74**, 560–564 (1977).

26. Saiki, R., Gelfand, D.H., Stoffel, S., Scharf, S.J., Higuichi, R., Horn, G.T., Mullis, K.B., and Ehrlich, H.A.: "Primer-directed enzymatic amplification of DNA with a thermostable DNA polymerase." *Science* **239**, 487–491 (1987).

27. A comprehensive and clear presentation of the history of cellular molecular biology can be found in Darnell, J., Lodish, H., and Baltimore, D.: *Molecular Cell Biology,* second edition, Scientific American Books, 1990.

28. It is important to realize that not only is all *structural* information contained in the DNA of the cell nucleus, but also all the particular sequences by which this information is accessed in the course of development of an organism

and then later blocked. Thus the DNA also contains a *temporal* dimension. The process of differentiation of cells and organs is connected with particular patterns of gene activation. For example, the extent to which the specific DNA configuration of a liver or brain cell is still reversible is a matter of some debate. Cloning experiments in which somatic cells were fused with egg cells whose DNA had been previously removed have led to the supposition that there might be one or more steps in the development of cells in which the "totipotency" of a somatic cell is lost. It is sacrificed, so to speak, on the altar of "specialization." The recently published experiment on the cloning of a sheep from an udder cell shows, however, that totipotency can continue to exist in certain differentiated cells. See Schnieke, W. et al., *Nature* **385**, 810–813 (1997).

29. See also Drews, J.: *Wissenschaftliche Paradigmen in der Medizin.* Editiones Roche, Basel 1992.

30. For this purpose we may pursue two principal paths. The one path leads from gene to protein, the other from protein to gene. Whoever would take the first path must first isolate mRNA from particular cells such as liver or brain. This RNA is then transcribed into cDNA by reverse transcriptase and afterwards converted by DNA polymerase into two-stranded DNA. These DNA fragments are then inserted into vectors. With the vectors thus treated, host cells (bacteria, yeast cells) are transfected in such a way that each cell may take up one DNA molecule. One then has a cDNA library. The individual bacteria (each contains a specific DNA derived from one mRNA) can be cultured on agar in such a way that one cell produces in turn a colony that contains a particular type of transfected DNA molecule. This DNA can then be sequenced. Peptides can be synthesized corresponding to the nucleotide sequence, which can then be used for immunization of animals and for obtaining antibodies. With the help of antibodies obtained in this way, coded proteins, coded by the corresponding genes, can be isolated. The second way, leading from protein to gene, is sketched in the text.

31. Köhler, G. und C. Milstein: "Continuous cultures of fused cells secreting antibodies of predefined specificity." *Nature* **256**, 495 (1975).

32. This four-phase model is extensively described in Drews, J.: "Medicine and Genetic Engineering: just another method or a new paradigm?" *Arzneimittelforschung/Drug Research* **41** (1) 1, 94–100 (1991).

33. The information reported here can be found in *Human Genome News* 7, 3 and 4 (1995).

34. Dietrich, W.F. et al.: *Nature* **380**, 149–152 (1996). See also Jordan, E. and Collins, F.S.: "A march of genetic maps." *Nature* **380**, 111–112 (1996).

35. Jordan, E. and Collins, F.S.: "A march of genetic maps" and Patricia Kahn: "Human genome projects. Sequencers split over data release." *Science* **271**, p. 1799 (1996).

36. Oliver, St. G.: "From DNA sequence to biological function." *Nature*, **379**, 597–600 (1996).

37. Oliver, St. G.: "From DNA sequence to biological function." p. 598.

38. Oliver, St. G.: "From DNA sequence to biological function." pp. 598–599.

39. "From fly to man, cells obey same signal." Science Times. *New York Times*, Tuesday, January 5, 1996, B5, and Rubin, G.A.C.: "Spreading genetic transformation of drosophila with transposable element vectors." *Science* **218**, 348–353 (1982).

40. Oliver, St. G.: "From DNA sequence to biological function." p. 599.

41. Soluble proteins can today be identified in large data banks by sequence analyses, which show the presence of a signal peptide. An experimental method for establishing the presence of genes that code for soluble proteins operates by the gene-trap method. Unknown DNA transcripts obtained from cDNA are inserted into yeast vectors that contain an invertase gene. With such a construct, transfected invertase-negative yeast cells grow, because the transfected signal peptide gives them the possibility to excrete the invertase contained in the vector. Subsequently, they split sucrose, and grow. If a gene that contains no signal peptides is transfected, the invertase gene remains in the cell and cannot be put to use for the splitting of sucrose: The cell cannot grow.

42. A number of relevant papers are cited in Darnell, J., Lodish, H., and Baltimore, D.: *Molecular Cell Biology*, 2nd edition, p. 187 (1990).

43. Summarized in Sokol, D. und Gewirtz, A.: "Gene therapy: basic concepts and recent advances." Crit. Rev. In: *Eukaryotic Gene Expr.* **6** (1), 29–57 (1996).

44. Bresch, C.: *Klassische und molekulare Genetik*. Springer, Berlin 1964.

45. Overviews of the various vectors can be found in, for example, Schimada, T.: *Acta Paediatrica Japonica* **33**, 176–181 (1996); Brenner, M.K.: "Human somatic gene therapy: progress and problems." *Journal Int. Med.* **237**, 229–239 (1995); Crystal, R.: "Transfer of genes to humans: early lessons and obstacles." *Science* **270**, 404–410 (1995).

46. "T-Lymphocyte-directed gene therapy for ADA-SCID: Initial trial results after 4 years." Science **270**, 475–480 (1995).

47. There are indications from colleagues at the firm Somatix concerning several patients with melanomas who seem to have reacted dramatically to the gene-therapeutic principle of treatment we have described. In addition, cases with renal adenocarcinoma responded well to this treatment. However, the number of patients involved is still too small for reliable conclusions to be drawn.

48. See Crystal, R., *Science* **270**, p. 404 (1995). Grossman, M. et al.: "A pilot study of ex vivo gene therapy for homozygous familial hypercholesteremia." *Nature Medicine* **I**, 1148–1154 (1995).

49. See Crystal, R., *Science* **270**, p. 404 (1995).

# Chapter 5

1. Jackson, E.K. und Garrison, J.C.: "Renin and Angiotensin" in Goodman and Gilman's *The Pharmacological Basis of Therapeutics*, 9th edition. McGraw Hill, New York 1996, pp. 751 *ff*.

2. Jackson, E.K. und Garrison, J.C.: "Renin and Angiotensin." pp. 733 *ff*.

3. Cushman, D.W., Cheung, H.S., Sato, E.F., and Ondetti, M.A.: "Design of potent competitive inhibitors of angiotensin-converting enzyme. Carboxyalkanoyl - and mercaptoalkanoyl amino acids. *Biochemistry* **16**, 5484–5491 (1977).

4. Usually, the first blind screening would be carried out by adding the substance to be tested to a liquid culture of microorganisms (final concentration between 1 and 100 μg/ml) and then measuring the density of the bacteria after several hours' incubation. Subsequent to these preliminary, orienting, experiments, active substances were tested in greater dilution. In this way the minimal inhibitor concentration (MIC) was obtained, that is, the smallest concentration of the substance under examination that still was able to achieve a complete suppression of the growth of the microorganism in question. The level of the MIC depends on various parameters, among which are the composition of the medium in which it is tested, the oxygen tension, the pH, and the temperature of incubation.

5. See: Maxwell, R. and Eckhardt, S.B.: *Drug Discovery, a Casebook and Analysis.* Humana Press, Clifton, N.J., 1990.

6. A very understandable presentation on combinatorial synthesis, libraries, screening strategies, and future prospects for this technology can be found in Gordon, E. et al.: "Applications of combinatorial technologies to drug discovery." *J. Med. Chem.* **37**, 1385–1401 (1994). See also Appel, K.C. et al.: "Biological screening of a large combinatorial library." *J. Biomol. Screening* **I**, 27–31 (1996). A report on optimizing candidate substances by a "genetic algorithm" is given in the following paper: Weber, L. et al.: Optimization of the biological activity of combinatorial compound libraries by a genetic algorithm. *Angewandte Chemie. Int. Engl. Ed.* **34**, 2280–2282 (1995). An extensive presentation that goes especially into the problems of discovering new drugs can be found in Terret, N. et al.: "Combinatorial synthesis-the design of compound libraries and their application to drug discovery." *Tetrahedron* **51**, 8135–8173 (1995). Chemical methods of combinatorial synthesis have been extended recently through recombinant methods. The principle of these techniques is based on elucidating the genetic and biochemical bases of the synthesis of secondary metabolites, primarily the synthesis of polyketides. This class of substances comprises molecules with very different chemical properties, such as macrolides, rifamycins, tetracyclines, doxorubicin, rapamycin, lovastatin, and many others. What these structurally so diverse compounds have in common is the principle of chemical synthesis by which they arise. The synthesis of these structures is accomplished with large proteins containing several linearly ordered enzyme activities. The synthesis process begins with an acyltransferase, an acyl carrier protein, a ketosynthetase, further acyltransferases, ketoreductases, dehydrogenases, and further "carbon-modifying" enzymes. These enzyme activities are arranged on one or more proteins as in the functioning of an assembly line. The synthesis of a polyketide, for example a macrolide, begins at one end of the polyenzyme, and the product is built up step by step, being sent down the assembly line until finally, at the end of the line, a finished product is split off. The genes that code for the enzyme are ordered in the same way on the genome of microorganisms as the enzyme activities on the protein. Through inactivation or the exchange of individual enzymatic activities, the

order or the nature of the synthesis steps on the "assembly line" can change. In this way one can create completely new kinds of substances with potentially new effects. The combinatorial potential of polyketide synthesis amounts to several hundred thousand to several million molecules. It can be represented by the following formula:

$$CP = AT^L \times [AT^E \times 4]^M$$

CP = combinatorial potential

$AT^L$ = Acyltransferases that catalyze the initial step of the synthesis

$AT^E$ = Number of chain-lengthening malonyl transferases

4: = The number of carbon-modifying enzymes

M = The number of successively connected modules, each of which is made up of several enzyme activities

A typical example: $CP = 1 \times [2 \times 4]^7 = 2$ million.

The potential of a genetically controlled recombinant chemistry of microorganisms is recognizable in outline. On the process of discovering new drugs it has had as yet no influence. Since polyketides have a prominent place among the drugs used frequently today, it is possible that the recombination of these structures will likewise lead to a large number of new and valuable substances. For further reading, see Verdine, G.L.: "Combinatorial chemistry of nature." *Nature* **384**, 11–13 (1996). Hutchinson, R.C.: "Drug synthesis by genetically engineered microorganisms." *Biotechnology* **12**, 375–380 (1994). McDaniel, R. et al.: "Rational design of aromatic polyketide natural products by recombinant assembly of enzymatic subunits." *Nature* **375**, 549–554 (1995).

7. The system introduced by the Genetic Institute for identifying soluble proteins has been described in a variety of publications in generally understandable form. For example, "The haystack gets smaller." *Business Week*, October 21, 1996. Potera, C.: "Genetic Institute unveils a new platform for gene isolation and functional analysis." *Gen. Eng. News* **16** (18) (15 October 1996). A more comprehensive description can be found in Erickson, D. "GI goes fishing." *In vivo* **14** (9) (1996).

8. The reader who is interested in studying the individual steps of drug development is referred to the following books: *Grundlagen der Arzneimitteltherapie.* W. Dölle, B. Müller-Oerlinghausen, and U. Schwabe, eds. BI Wissenschaftsverlag, Mannheim, Wien, Zürich 1986. The series *Drugs and the Pharmaceutical Sciences.* James Swarbrick, ed., vols. 1–78. Marcel Dekker Inc., New York, Basel, Hong Kong 1996 and earlier.

9. See also: Banker, G.: "Drug products: Their role in the treatment of disease, their quality, and their status as drug delivery systems." In: *Modern Pharmaceutics*, G. Banker and Ch. Rhodes eds. 3rd. ed. Marcel Dekker, New York 1996.

10. A. Evans gives a clear and comprehensive list of toxicity tests in "New drug approval in the U.S." Chapter 2 of *Nonclinical Drug Testing.* Parexel (1995).

11. The attitude of the FDA to "surrogate markers" has changed considerably over the course of the AIDS epidemic.

12. The validity of viral load as a "surrogate marker" is treated in Deyton, L.: "Importance of surrogate markers in evaluation of antiviral therapy for HIV infection." *JAMA* **276**, 159–160 (1996). See also Levy, J.A.: "Is there truth in numbers?" *JAMA* **276**, 161–162 and Mellors, J. et al.: "Quantification of HIV RNA in plasma predicts outcome after seroconversion." *Ann. Int. Med.* **122**, 573–579 (1995).

13. Yong Ming Li et al.: "Prevention of cardiovascular and renal pathology of aging by the advanced glycation inhibitor aminoguanidine." *Proc. Natl. Acad. Sci.* **93**, 3902–3907 (1996). Also Altheon Comp., Ramsey, N.J., personal communication.

14. Di Masi, J. et al.: *Research and Development costs for new drugs by therapeutic category.* PharmaEconomics 1995.

15. Mathieu, M. et al.: *New drug approval in the United States.* Parexel 1995.

16. *New drug approval in the United States*, p. 91.

17. Shulman, S. and K. Kaitin: "The prescription drug user fee act of 1992. A five year experiment for the industry and for the FDA." *PharmacoEconomics*, 9 Feb. 1996, pp. 121–133.

18. Mathieu, M. et al.: *New Drug Approval in the United States*, pp. 103–104.

19. Shulman, S. and Brown, J.S.: "The Food and Drug Administration's early access and fast track approval initiatives: how have they worked?" *Food and Drug Law Journal* **50**, 503–531 (1995); see particularly pp. 515– 517.

20. Evers, P.T. et al.: *New Drug Approval in the European Union.* Parexel 1995.

21. Mathieu, M. et al.: *New Drug Approval in the United States*, pp. 111 *ff.*

22. What follows is based to a great degree on Nielsen, R.: *Handbook of Federal Drug Law.* Lea und Febiger, 2nd edition, Philadelphia 1992.

23. Nielsen, R.: *Handbook of Federal Drug Law.*

24. Nielsen, R.: *Handbook of Federal Drug Law*, p. 8; and from the European viewpoint, Drews, J.: "Orphan drugs aus europäischer Sicht." *Die Pharmazeutische Industrie* **50**, 803–805 (1988).

25. What follows is based on the following sources: Stewart, Ronald B.: *Tragedies from Drug Therapy*, chapter III. Charles Thomas, Springfield, Ill. 1985, and McBride, William: "Thalidomide embryopathy." *Teratology* **16**, 79–82 (1977). Also Knightley, P., Evans, H., Potter, E., and Wallace, M.: *Suffer the Children: The Story of Thalidomide.* Viking Press 1979, chapter. 2: "When Is a Rat Asleep?"

26. See also Lenz, W. and Knapp, K.: "Foetal malformations due to thalidomide." *German Medical Monthly* VII, 200–206 (1962). This is the English edition of the *Deutsche Medizinische Wochenschrift.*

27. Delahunt, C.S. and Lassen, L.J.: "Thalidomide syndrome in monkeys." *Science* **146**, 1300–1305 (1964). Lucey, J. and Behrmann, R.: "Thalidomide: effect upon pregnancy in the Rhesus monkey." *Science* **139**, 1295–1296 (1963).

28. Mossinghoff, G.J.: "Health care reform and pharmaceutical innovation." *Drug Information Journal* **29**, 1077–1090 (1995). Schwartz, H.: "Why research costs are rising so rapidly—the Schwartz view." *Scrip*, 15–16, March 1997.

29. Mathieu, M. et al.: *New drug approval in the United States.* Parexel, 1995.

30. Hjalmarson, A. und Olsson, G.: Myocardial infarction—effects of β-blockade." *Circulation* **84**, suppl VI, 101–107, 1991.

## Chapter 6

1. See also Drews, J.: "Research in the pharmaceutical industry." *Eur. Management J.* **7**, 23–30 (1989).

2. We (Stefan Ryser und Jürgen Drews) examined in 1993 the question of how gene-therapeutic methods would affect drug therapy if all gene-therapeutic approaches undertaken in that year turned out to be successful. At that time we came to the conclusion that in such a case drug sales to the tune of twelve billion dollars would become obsolete. Based on the pharmaceutical market in 1993, this represented about seventeen percent of drug sales worldwide.

3. See: Drews, J.: "Science and technology are the prime movers of the pharmaceutical industry." *Chimica oggi/chemistry today* **12**, 9–13 (1994).

4. This possibility is not disputed by the research directors of the major pharmaceutical firms. The author's opinion is not, however, shared by all research directors. Many are of the opinion that the industry needs to attract a whole new type of scientist, one to which the behavior patterns described here do not apply. They operate on the assumption that the drug discovery process will develop into a largely automated, miniaturized, and computer-supported process. For such a process, so goes the argument, a completely different type of researcher will be necessary, one who is characterized by outstanding knowledge of information systems and optimization and who is motivated almost exclusively by predetermined goals (new medicines). This type reacts more positively to an environment characterized by interdisciplinarity and business acumen than did his "scientifically" motivated predecessor.

5. The founding of "Roche Bioscience" on the campus of the former firm Syntex, in Palo Alto, California, which was taken over by Roche, obeys these principles. The thoroughly positive experience with this organizational arrangement has encouraged Roche to apply the same basic principles to its traditional research centers.

6. This point of view is not given sufficient attention in the industry. Scientists who enter the industry often follow their own scientific interests, although in many cases it must be clear to them that they hardly have a chance of discovering a new medicine. And businesspeople often think that they can expect greater results if they simply invest more money in a project or field of work. But this connection exists only within precisely defined scientific and technological boundaries.

7. The relevant details can be found in the book by Robert A. Maxwell und S.B. Eckhardt: *Drug Discovery.* Humana Press, Clifton, N.J. (1990).

8. Wells, J. and deVos, A.M.: "Hematopoetic receptor complexes." *Ann. Rev. Biochem.* **65**, 609–634 (1996). See in particular pp. 624 *ff.* Wells und deVos point out that a small molecule must be capable of inhibiting the interaction of two

large proteins with a relatively flat interaction plane, since the binding energy doesn't extend evenly across the whole surface of such an interaction but can be concentrated over various narrowly circumscribed areas. The "disturbance" of one such area could inhibit the interaction of the proteins among themselves. In practice, however, there are hardly any substances that make use of this situation. Clackson, T. and Wells, J.: *Science* **267**, 383–386 (1995) and Cunningham, B.C. and Wells, J.: *Journal of Molecular Biology* **234**, 554–563 (1993).

9. In a development process determined more and more by financial analysis, the implementation of this principle has often become difficult.

10. Though this peer review process has not been institutionalized in all firms, nevertheless, an efficient system of external consultants often sees to the "objective" evaluation of a firm's projects. To this extent, the companies are united by a network of academic consultants. By means of this network, mechanisms of competition become activated. Information is, to be sure, not provided directly, and the names of other companies are not revealed. However, a "well advised" firm often learns from the opinion of the expert evaluator whether or not it is competitively well positioned in a particular area.

11. The validity of this information can be checked in detail. For this the following books are particularly useful: Sneader, W.: *Drug Discovery, The evolution of modern medicines*. John Wiley and Sons, New York (1985), as well as *Discoveries in Pharmacology*. 2 vols. M.J. Panham und J. Bruinvels, eds., Elsevier, Amsterdam, New York, Oxford (1983). It is nonetheless worthy of note that there is a movement in this age of genome research and combinatorial chemistry to return again to blind screening—though to be sure on a higher technical level. However, it remains to be seen whether the transformation of drug discovery into a "drug discovery process" will have the success that was denied to conventional blind screening.

12. Under certain conditions the functional interpretation of the genome can be simplified by a study of the proteome. By means of two-dimensional gel electrophoresis, proteins can be separated cleanly and reproducibly. With this technique it is possible, for example, to determine which proteins are phosphorylated under certain conditions, for example after binding of a ligand by a receptor. By means of a partial sequencing of such proteins one can be referred back to the genome and recognize the function of the genes within regulatory circuits or signal pathways.

13. In this regard see the articles from the laboratory of Stuart Schreiber; for example, Chen, K. James and Schreiber, S.: "Combinatorial synthesis and multi-dimensional NMR spectroscopy: an approach to understanding protein–ligand interactions." *Angew. Chemie* (international edition) **34**, 953–963 (1995); and Belshaw, P. et al: "Controlling protein association and subcellular localization with a synthetic ligand that induces heterodimerization of proteins." *Proc. Natl. Acad. Sci.* **93**, 4604–4607 (1996).

14. Research directors of large firms see, at least for now, expending 30% of the research budget for joint ventures and research outside the firm as a sort of upper limit. A significant influence on this limit is the management time that

must be spent to ensure that these joint ventures function, that technologies are really assimilated, and that concrete projects for in-house research actually result.

15. Managers who are themselves not physicians or scientists often act in ways that are counterproductive in the establishment of independent research units. They overlook the fact that research today is an interdisciplinary process that extends across institutional borders and that synergistic effects can result from the collaboration of the firm's own research centers. Often it is a matter of something truly banal, such as the exchange of reagents, methodological assistance, or help in finding another laboratory in which specific assistance can be obtained. In collaborations with third parties as well a certain degree of coordination—even if only informing the relevant parties of the fact—are desirable. The author has often experienced that a licensing team of a major affiliate of the company pays a visit to another country (Japan, for example) without anyone in the central office knowing anything about it. They become aware of the fact only when a group from the home office visits the same partner. Then they realize that a team of their own company has already negotiated with the Japanese partner. Not a very efficient way of dealing with collaboration or licensing!

16. Schumpeter, J.A.: *Business Cycles. A Theoretical, Historical and Statistical Analysis of the Capitalist Process*. McGraw-Hill, New York 1939.

17. Drews, J.: An innovation strategy for the pharmaceutical industry. *Drug News and Perspectives* **7** (3), 133–137 (1994).

18. Many of them can be found in Maxwell, R.A. and Eckhardt, S.B.: *Drug Discovery: A Casebook and Analysis*. Clifton, Humana Press 1990.

19. Report of the National Institutes of Health: *The biomedical research and development price index*. Nat. Inst. of Health, Bethesda, MD, USA (1996).

20. To be sure, there are also limitations. Often "gene knockouts" demonstrate no unambiguous phenotype. Examples are the inactivation of the prion gene or of the gene for the β-amyloid precursor protein. These results could be indications about the strong, functional interdependence of various gene products, with the result that functional deficits hardly occur if one of these genes is inactivated. Conversely, transgenic animals often show very pleiotropic effects. Here, too, functional interdependence could be coming into play. The medical or pharmacological evaluation of pleiotropes with genetic effects is still in its infancy.

21. According to the Pharmaceutical Research Based Manufacturer's Association (PhRMA), 1996.

22. This development, which seems to be continuing, is disturbing also for another reason. Let us suppose that the research and development costs for a new preparation came to $500 million. If such a sum earned ten percent interest, that is, a rate easily attainable by large companies if they put their capital to work in the international financial markets, then after ten years they would receive more than 1.4 billion dollars. Since research and development are paid for in pretax earnings and these lie in the range of twenty to thirty percent of sales, then roughly speaking, a drug that costs $500 million to develop would have to generate total sales on the order of $7 billion to cover its costs. This means sales of

$700 million per year over ten years simply to recoup the investment costs. Of course, the life span of a successful drug is longer than ten years. In our example, the drug would start earning money only after ten years. What medicines and what indications fulfill this requirement? Only new substances that can affect frequent, serious diseases come at all under consideration. Medicines for many serious and not at all rare diseases would hardly have a chance of being developed under these circumstances. If development costs cannot be drastically reduced—and this can happen only through a fundamental revision of the process—the industry will gradually withdraw from its own innate areas of work. See also Chapter 7.

## Chapter 7

1. German statistical yearbooks 1992–1996.

2. According to the National Institutes of Health. See the *Biomedical Research and Development Price Index* (1996).

3. Drug costs as a percentage of healthcare costs in 1993 were 8.5% in the USA, 15.5% in the UK, 18.6% in Germany, 11.5% in the Netherlands, 14.7% in Canada, 16.3% in France, and 11.6% in Italy. Source: www.pharma.org/facts/industry/chapter4/htm.

4. A critical analysis by David Blumenthal on this theme has appeared in several issues of the *New England Journal of Medicine*: "Quality of care. What is it?" *New Engl. J. Med.* **335**, 891–893 (1996); "Measuring quality of care." *ibid.* **335**, 966–969, (1996); "Improving the quality of care." *ibid.* **335**, 1060–1062; "The origins of the quality of care debate." *ibid.* **335**, 1146–1148.

5. This study was published in two places. A complete description of the methods can be found in Drews, J. and Ryser, S.: *Drug Information Journal* **30**, 97–108 (1996). A summary and strategic evaluation is in Drews, J.: "The impact of cost containment on pharmaceutical research and development." Tenth Centre of Medicines Annual Lecture, June 1995 (this can be ordered from CMR, Woodmanssterne Rd., Carshalton, Surrey, SM5 4DS, UK. Fax: +44 181-770 7958).

6. With a somewhat different approach, Jan Leschly, chief executive officer of SmithKline und Beecham, came to similar conclusions.

7. See also reference 5.

8. Drews, J.: "Die Führung von Mitarbeitern in Innovationsprozessen." *NZZ* **211**, 123, p. 65, 1990.

9. Brunt, J.: *BioWorld Financial Watch*, Jan. 22, 1996.

10. In this regard see also footnote 5.

11. "Public funding for basic biomedical research," *Science* **274**, 491 (25 October 1996).

12. Blumenthal, D. et al.: "Relationships between academic institutions and industry in the life sciences—an industrial survey." *New Engl. J. Med.* **334**, 368–373 (1996).

13. Some American medical schools wish to start their own firms for exploiting research that is conducted within their confines. These companies would be charged with either selling results of university research directly to industry or

developing them further at their own cost, after which they would be sold or licensed to industry. The university researchers involved would be stockholders in such firms, and so they, too, would profit from the commercial exploitation of their own ideas and discoveries. These "exploitation companies" should be directed by independent management. According to one model, the university would own a portion of the firm, but would not become the majority stockholder. A similar idea was presented in the context of the bioregional competition of the Heidelberg Work Group (proposal of the Bioregion Rhein–Neckar Triangle to the BioRegio-Wettbewerb 1996).

14. In this regard, see the annual report of the consulting firm Ernst and Young on the status of the biotechnology industry in the United States and Europe.

15. The geographical distribution of these firms stands in inverse proportion to the degree of resistance that has been put up in the various European countries against the development of biotechnology. England (UK) has the great majority of such firms, followed by France and Benelux. Germany languished for a time at the bottom of the list. Only recently has there been a change in its status. Some of the conditions that determine where biotechnology firms will be established are the following: the availability of capital, above all venture capital; low taxes; financial incentives for employees; a good scientific milieu that shows an interest in entrepreneurial activity; a well-disposed public and a positive regulatory climate. In the German-speaking countries of Europe all of these factors have been lacking for a long time. As a result, Germany has fallen behind the United States so significantly that there is no chance of it catching up in the foreseeable future. Through concentrated efforts it ought to be possible for Germany to at least reach the level of France or the UK within a few years. See also the documents presented at the "BioRegio" competition in Germany (available from the Bundesministerium für Forschung und Technik).

16. See also Drews, J.: "The changing roles of industry and academia." *Scrip* magazine, June 1993.

17. See also Drews, J.: "Strategic choices facing the pharmaceutical industry: a case for innovation." *Drug Discovery Today* 2, 72–78 (1997).

18. Drews, J.: *Drug Discovery Today* 2, 72–78 (1997) and "An innovation strategy for the pharmaceutical industry, *Drug News and Perspectives* 7 (3), 133–137 (1994).

19. Fish, D. et al.: "Development of resistance during antimicrobial therapy: A review of antibiotic classes and patient characteristics in 173 studies." *Pharmacotherapy* 15 (3): 279–291 (1995). Cohen, M.L.: "Epidemiology of drug resistance: implications for a post-antimicrobial era." *Science* 257, 1050–1055 (1992). See also the editorial "The microbial wars" by Daniel E. Koshland, also in *Science* 257, 21 August 1992, and Neu, H.C.: "The crisis in antibiotic resistance." *Science* 257, 1064–1073 (1992).

20. A brief description of the various origins of science and technology can be found in Drews, J.: "Erkennen oder Handeln? Wissenschaftsverständnis im Wandel," in: *Forschung bei Roche*, J. Drews und F. Melchers, eds., Editiones Roche

1989. A more extensive treatment can be found in Böhme, G. et al.: *Starnberger Studien: Die gesellschaftliche Orientierung des wissenschaftlichen Fortschritts.* Edition Suhrkamp, Frankfurt/Main 1978.

21. Böhme, G. and W. v.d. Daele, "Die Verwissenschaftlichung von Technologie," in *Starnberger Studien,* pp. 339–375 (1978).

22. See Chapter 2.

23. See also Drews, J., *Scrip* magazine, June 1993.

24. Genentech is a typical example. From such "free time" a number of development substances have emerged over the years.

# ⩓⩓⩓ Glossary

**Abscess**   Bacterial infection with suppurating colliquation

**ACE-Inhibitors**   (Angiotensin Converting Enzyme Inhibitors) substances that inhibit the transformation of Angiotensin I into the more active Angiotensin II

**Acetolamid**   A diuretic medicine

**Adrenergic**   Substances or receptors that produce or convey epinephrine-like effects

**Adrenotropic**   reacting to the adrenal hormones epinephrine or norepinephrine

**Agonist**   A substance that triggers a positive (usually excitatory) effect at a receptor

**Amnesia**   Loss of memory

**Amphetamine**   A central nervous system stimulant

**Analgesia, Analgesic**   Pain-inhibiting or -reducing measures and drugs for such purposes

**Anemia**   Pathological deficiency in red blood cells and/or hemoglobin, the oxygen-carrying component of the blood

**Anesthesia**   Total or partial loss of sensation

**Antagonist**   A substance that inhibits an agonistic effect

**Antifungal**    An agent employed against a fungus, e.g., *antifungal* chemotherapy

**Antisepsis**    Destruction of disease-causing microorganisms to prevent infection, through spraying of carbolic acid (phenol)

**Antitussive**    A cough suppressant

**Arrhythmia**    Irregular heartbeat

**β-Lactams**    Antibiotics that contain a β-lactam ring. These include penicillins and cephalosporins

**Benzodiazepines**    Sedatives with a particular structure

**β-Blocker**    Substances that block the β-receptors. Suppressors of particular effects of epinephrine

**Bioassay**    Determination of the strength or biological activity of a substance based on the reactions of cells (as opposed to purely biochemical tests)

**Blood plasma**    Uncoagulated blood freed from all cellular components

**Bradycardia**    slow heartbeat

**Calcium antagonists**    Substances that inhibit the influx of calcium through particular calcium channels

**Carbonic anhydrase**    An enzyme that releases hydrogen ($H^+$) and bicarbonate ($HCO_3^-$) from carbonic acid

**Centimorgan**    Length of genetic material within which the exchange of genetic material (crossovers) occurs in only one percent of cases. In man this corresponds to one million base pairs

**Chrysoidin**    A dye, chemically related to prontosil, the first sulfonamide

**Cloning**    Isolation and replication of genes

**Colony stimulating factors**    Hormones that stimulate the creation of new blood cells

**Cystic fibrosis**    Hereditary disease in which due to a defective gene for an ion channel only very viscous mucus can be produced. Patients suffer from occlusion of the small bronchi, leading to infections and difficulty in breathing

**Cytoskeleton** The internal framework of a cell, composed largely of actin filaments and microtubules, providing both structure and mobility to the cell

**Dialysis** Cleansing of the blood (in cases of kidney failure)

**Dilatator pupillae** The muscle that enlarges the pupil

**Diploid** The number of chromosomes in a normal (somatic) cell

**Diuretic** A substance or drug that tends to increase the discharge of urine

**DNA** Deoxyribonucleic acid—genetic material

**DNase** An enzyme that catalyzes the hydrolysis of DNA

**Drug delivery** Measures to ensure that a drug finds its way to the site of its action

**Encoding** Information (in RNA) for building a protein

**Endothelin** A hormone that constricts the blood vessels in case of injuries. Endothelin antagonists are substances that dilate blood vessels

**Ephedra** A genus of plants that contain ephedrine

**Epidemiology** The study of the frequency and distribution of diseases

**Episomal** Pertaining to genetic information that is not necessarily contained in the chromosomes

**Erysipelas** A streptococcus infection of the skin and subcutaneous tissue

**Erythropoietin** A substance that stimulates the production of red blood cells by bone marrow

**Eukaryote** Cells that contain a distinct membrane-bound nucleus (as opposed to bacteria or prokaryotes)

**Expression** Activation of genes in certain cells or organisms

**FDA** Food and Drug Administration—the American registration agency for drugs

**Filariasis** A tropical disease caused by nematode worms

**Furosemide**   A diuretic

**G-CSF**   A colony stimulating factor that stimulates the production of white blood cells

**G-Protein**   Element of an intracellular signal transfer that functions with the ligands GTP (guanosine-5-triphosphate)

**Gangrene**   Death and decay of body tissue

**Glomerulosa**   Layer of the renal cortex in which the hormone aldosterone is produced and secreted

**GM-CSF**   A colony stimulating factor that stimulates the production of granulocytes and monocytes

**Gram-positive, -negative**   Relating to a bacterium that retains the violet stain used in the method of Hans Gram (Danish physician 1853–1938). Used in the classification of bacteria

**Haploid**   Pertains to the chromosome number in germ cells, which contain only half the normal diploid chromosome set

**Hematology**   The study of blood and diseases of the blood

**Hepatitis**   Viral infection of the liver. Most frequent forms: Hepatitis A, Hepatitis B, and Hepatitis C, the last two of which are spread by contact with blood

**Histology**   The study of tissue, concerned with microscopic and submicroscopic tissue structure

**Humoral**   Relating to bodily fluids

**Hydrochlorothiazide**   A diuretic drug that inhibits the reabsorption of sodium in the distal renal tubules

**Hypercholesterolemia**   An excess of cholesterol in the blood. It is a risk factor for cardiovascular diseases

**Hypertension**   High blood pressure

**Hypnotic**   A sleep-inducing drug

**IND**   Investigational New Drug Application. Application for the first clinical trial of a new drug

**Indication**   A indicator that particular therapeutic measures should be taken, e.g., a disease

**Infestation**   Infection by parasites

**In situ**   Found in its natural place

**Integrins**   Proteins occurring on the surface of many cells that determine the cohesion of cells in tissues or effect the attachment of one cell type to another

**Interleukin-3**   Colony stimulating factor with a broad spectrum of activity

**Ion channel**   Opening in the cell membrane through which electrically charged particles (anions or cations) can enter or leave the cell

**Ipecacuanha**   Plant whose roots contain emetine

**Isomers**   Compounds composed of the same elements in the same proportions but with different arrangements of atoms and therefore with different properties

**Macrolide**   A chemical structure. Antibiotics with this structure have a primarily Gram-positive spectrum of activity

**Miasma**   A poisonous atmosphere once thought to rise from swamps and putrid matter and cause disease

**Mutagen**   An agent that can cause genetic mutations

**National Institutes of Health (NIH)**   Federal medical research institute in the USA

**Neuroleptics**   Drugs used in the treatment of schizophrenia

**NMDA-Receptor**   N-Methyl-D-Aspartate-Receptor. Receptor in the brain that reacts to excitatory amino acids

**Nosology**   The study of classification of diseases

**Nuclear magnetic resonance (NMR)**   Diagnostic procedure based on the absorption of electromagnetic radiation of a specific frequency by an atomic nucleus placed in a strong magnetic field

**Percussion**   Determination of the location and extent of organs by external tapping or "percussion" of the body

**Phenothiazine**  A class of chemical substances used in the treatment of psychosis

**Phenotype**  The observable characteristics of an organism (as opposed to the genotype), the total of inherited characteristics

**Polyarthritis**  chronic inflammation of several joints

**Portfolio**  In the pharmaceutical industry, the totality of all drugs in research or development (research and development portfolios)

**Positron emission tomography**  A technique that provides a computer-generated image of a biological activity

**Prevalence**  The frequency of a disease in a population at a particular time

**Promoter**  A DNA molecule to which RNA polymerase binds, initiating the transcription of messenger RNA

**Protease inhibitor**  Inhibitor of proteases, for example, ACE-inhibitor, inhibitor of viral proteins; that is, antiviral substances

**Psychosis**  Mental disease characterized by loss of contact with reality

**racemic mixture**  Mixture of the optically left- and right-handed forms of a substance

**Receptor**  Binding site for particular molecules

**Recombinant proteins**  Proteins obtained by genetic recombination

**Reflex tachycardia**  Rapid heartbeat caused by lowering of blood pressure

**Remission**  Disappearance of symptoms of tumors after treatment

**Renal**  Pertaining to the kidneys

**Renal adenocarcinoma**  A malignant tumor of the kidney

**Reverse transcriptase**  A polymerase in retroviruses by which the RNA of the virus is transcribed into DNA

**Segregation**  Separation of paired alleles by cellular division or from one generation to the next

**Serendipity**   The faculty of making fortunate discoveries by accident. The word comes from a story by Hugh Walpole, in which three princes of Serendip find many things that they were not seeking

**Spasmodic**   Having the character of a spasm; convulsive

**Subarachnoidal bleeding**   Bleeding in the meninges through the rupture of a blood vessel

**Sympathicomimetic**   *see* adrenergic

**Teratogenic**   causing malformations

**Tetracycline**   Broad-spectrum antibiotic of typical structure, containing four aromatic rings

**Thrombolysis**   Dissolution of a thrombus

**Thrombosis**   Blood clot in a blood vessel

**TNF receptors**   Receptors for tumor necrosis factor (q.v.)

**Tomography**   Procedure by which the entire body or regions of the body are displayed in a sequence of closely spaced cross-sections

**Tranquilizer**   Sedative, usually containing benzodiazepine

**Transcription**   Here refers to the transcription of DNA into RNA

**Transcription factor**   A protein that activates the transcription of a gene after binding to a promoter

**Transfection**   Infiltration of DNA into living cells

**Transgene**   A gene infiltrated into the germline of another species

**Translation**   Here the translation of messenger RNA into protein

**Trypanosomes**   Parasitic flagellate protozoans transmitted to the vertebrate bloodstream, lymph, and spinal fluid by certain insects, causing diseases such as sleeping sickness and nagana

**Tumor necrosis factor (TNF-$\alpha$)**   A protein (cytokine) that has been shown experimentally to be capable of attacking and destroying cancerous tumors

**Vasoconstriction**    Constriction of a blood vessel

**Vasodilatation**    Dilation of a blood vessel

**Vasospasm**    Extreme constriction of a blood vessel, causing a reduction in blood flow

**Xenotransplantation**    Interspecies transplantation of organs, for example, the transplantation of an ape kidney into a human being

# Index